美术馆指南

基于设计的角度

A FIELD GUIDE TO CHINESE MUSEUM DESIGN

唐克扬　编著

Edited and Written by
Tang Keyang

生活·讀書·新知 三联书店

目录
Contents

城市
Urbanism

·········

关心美术馆就要关心美术馆的城市语境。大多数美术馆处于且属于城市，美术馆本身是城市文化的产物。中国城市罕见的发展机遇正是今天美术馆热的语境前提。美术馆并非单一为艺术家服务，应该是为城市生活做贡献的多功能公共空间。

·········

To consider a museum, one should consider its urban context. Most museums are in and of the city, a product of urban culture. The unusual opportunities offered by China's urban development are the context for Chinese museum mania. Museums are not merely spaces for artists but multifunctional public spaces, which contribute to the life of the city.

艺 术
Art

设计艺术展示空间怎能不理解艺术？ 艺术和美术馆设计都是没有标准定义和定则的，但是它们总还有一定的逻辑和标准来指导设计过程。一座好的建筑物未必就是一所好的美术馆，只有帮助推动当代的艺术观念的空间才是好的美术馆空间。

How could an architect design a space for the exhibition of art without understanding art? While art has no definite rules, and nor does museum design, there are no less certain logic and a sct of rationales which can guide the design process. A good building is not necessarily a good museum; only a space which promotes the concepts of contemporary art can be considered a good museum space.

设 计
Design

··········

建筑设计可以讨论吗? 这里的讨论一定是寻找共识而不是过分强调差异性。好的建筑设计是全方位的设计、切题的设计和在现有条件下追求最大效益的设计,而不一定是有个性的设计、新奇的设计和昂贵的设计。

··········

Is architectural design discussable? Discussions on this topic must seek consensus, instead of overly stressing differences. Good architectural designs are holistic, context-driven, and most effective and economic under current conditions, while not necessarily being idiosyncratic, newfangled or luxurious.

技　术
Technology

....

美术馆设计有自己的专业技术要求，当代的建筑学追求"生态效益"。 不管怎么说，一座能得到充分使用的美术馆就是绿色美术馆，不创造出额外的问题就是最好地解决问题，公众能够看得见的设计才是最好的设计。

....

Museum design has its own technical requirements. Contemporary architecture seeks "ecological viability". After all, a museum utilized to its full function is a green museum; requiring no extra solutions is the best solution; and a visible design is the best design for the public.

前言
Preface

我于2006–2015年担任中国美术馆馆长，期间，国家美术馆新建项目得到政府支持立项，我们着手招标设计，在这个过程中，我也得以了解和投入美术馆建筑设计的过程。我任职的近十年也可以说是中国的"美术馆的时代"，以"井喷"方式涌现的各种各样的新美术馆，既有各级政府作为公共文化建设项目推动的重点工程，也有私人和企业投入建设的特色空间，甚至还有一些难以用传统艺术史观念概括的"另类"美术馆。美术馆建设的实践迫切需要比照国际的经验，也需要批判性思想的注入，对于现状的反思和总结，以及适时跟进的专业知识的"充电"。

我深切地认识到一座美术馆的"建设"包含至少三个不同的层面：首先，在今天，一个新颖的艺术空间需要优异而富于创新的"硬件"，在这方面，空前发展的当代建筑学提供了前所未有的可能，各种新技术、新思想的撞击令人目不暇接；其次，这样的"硬件"又难以脱离它的功能而独立存在，美术馆原本是种单一的展示、宣教的空间，而今它已经成了当代文化生活中最活跃的一部分，兼有学习、娱乐、陶冶性情、社会交往的各种功能，它是一部文化生产的机器；最后，当代美术馆的"建设"将经历一个从理论到实践的漫长"实验期"，不像其他功能程序相对稳定的文化建筑，美术馆的"模型"并未成熟，有待当代的艺术实践和社会生活去检验，这既是向"下"看，即在中国情境中脚踏实地，又是向"前"看，毕竟，美术馆不是一类既已有之的事物，我们是在为中国艺术和中国文化的明天而建设。

经由2010年以来的国家美术馆新馆项目，我和几位老朋友——谢小凡副馆长、唐克扬博士结成了稳固和富有成效的工作关系。在这一过程中，我们遍访世界各大美术馆，和国内外最优秀的建筑师和艺术工作者深度切磋和磨砺。在充满挑战和收获的合作中，团队为美术馆建设工作积累了设计建造美术馆的宝贵的工作经验。他们的思想成果，今天终于成就了这一本小中见大的《美术馆指南》，这并非一份官方结论的"白皮书"，而是一份有分量的田野报告，凝聚了漫长时间以来的智慧、体会和思考。我向诸位读者诚挚地推荐这本书，希望它可以成为下一个阶段有关中国美术馆建设思想热潮的起点。

中央美术学院院长　原中国美术馆馆长

2016 年 6 月

I was the Director of National Art Museum of China (NAMOC) from 2006 to 2015. During my term, NAMOC's new building project was launched, with the support from the central government. We then started the international design competition, during which I got the opportunity to be acquainted with museum building design. The decade that I worked as the Museum Director can be called the "Age of the Museum" in China. New museums of all kinds have mushroomed: some are government-backed public cultural development projects; some are private or corporate-funded unique art spaces; others are "alternative" museums not even fitting into conventional art history concepts. Today, more than ever, it is imperative for Chinese museums to benchmark with international practice, as well as applying critical thinking, summarizing and reflecting on the state we are in, in order to continue to update our professional museum practice with time.

I am utterly convinced that "building" a museum involves at least three levels: first, an innovative art space nowadays requires well-built and innovative "hardware". In this regard, the unprecedented development of contemporary architecture has offered unparalleled possibilities, with an insurgence of new technologies and thinking; second, the "hardware" cannot stand isolated, with no consideration of the museum's functions. Art museums have shifted from the previous single-purpose exhibition and educational spaces into multi-functional spaces incorporating learning, recreation, cultivation and social interactions. They are cultural production machines, playing the most active role in contemporary cultural life; last but not the least, the development of contemporary art museums will undergo a long "experimental phase" from theory to practice. Unlike other cultural buildings with relatively mature functions and procedures, the art museum's immature "model" awaits to be developed and tested by contemporary art practice and social life. This requires looking "downward" with a down-to-earth approach that suits the China situation, as well as looking "forward" to the future. After all, art museums in China are not developed institutions. We are now building for tomorrow's art and culture.

Since the NAMOC new building project kicked off in 2010, I and my old friends – Mr. Xie Xiaofan, who was Deputy Director of NAMOC then, and Dr. Tang Keyang, have formed a committed and productive working team. During the project process, we have visited art museums around the world, and conducted in-depth communications and discussions with prominent architects and artists from China and abroad. Through challenging and rewarding cooperation for years, our team has accumulated valuable and precious experiences in designing and building art museums. The team's intellectual products have been compiled into this guidebook to museum design – a tiny book that illustrates big ideas. It is not an official whitepaper book, but a substantial fieldwork report that is a crystallization of the team's wisdom, experience and thinking. I highly recommend this book to readers and hope it will be the starting point of enlightenment on building art museums in China for the next stage.

Fan Di'an

President, Central Academy of Fine Arts

Director, National Art Museum of China (2006-2015)

2016. 6

为什么要出版《美术馆指南》

从建中央美术学院美术馆到承担国家美术馆新馆的筹建花了我十年的时间，这十年是学习的过程。所以，在2008年我把前一座美术馆的建设经验写成了一本书叫《展览美术馆》，它成了建馆者们的一本工作手册，二次重印，也都卖光了。这次建国家美术馆首先面临国际招标，于是写一本让外国建筑师能看明白的设计任务书就成了第一步。翻开可资借鉴的经验才觉得过去的想法太简单了，虽明白大致方向，但缺思考路径成了我当时的现实问题。好在范迪安馆长、朱青生教授一再向我推荐青年才俊唐克扬，他后来成为我们解决问题的有力助手。七年前，克扬刚从哈佛大学博士毕业回来，做了2010年威尼斯建筑双年展中国馆的策展人，有很多好的想法，带着经验和能力加入了我们这个团队，承担起编写设计任务书的具体任务。设计任务书的纲要是我草拟的，最初他还未加入。后来我去参加过一次高层会议，研究国家美术馆新馆建设之事，回来后在众多意见中我总结出四条会议要求：一是要体现中国元素，在中国为中国而设计，二是具有时代特征，三是实用大方，四是做到节能环保。如何把这四条融入设计任务书中，那就全靠克扬的本事。

朱熹讲求"格物致知"，但近百年中没能成为中国知识分子的主导，说大概、讲模糊，玄之又玄，成了做学问的风潮，而克扬不属这类，他的办法是细分又细分，务求理性与逻辑，把"美术馆"和"中国的"美术馆用各种方法进行剖析，把抽象的概念尽量变成具体的问题，从形态、色彩、空间、流线等方面阐述了我们想要的东西，从美术馆的哲学、功能、管理、建造等各方面，给建筑师明确的目标和方向，防止建筑设计不受任何约束的天马行空，这样，业主不知表达自己的诉求，由建筑师牵着业主的鼻子走的乱象才能有所纾解。

我们的招标分两阶段：第一阶段是概念性方案征集，放开让大家去设想，这一阶段沿用我写的招标大纲。第二阶段是在选出五名优胜者前提下进行邀请招标，这一阶段就使用了克扬主编的《国家美术馆建筑设计任务书》。它的出台不仅得到了各界的肯定，而且还得到了参赛选手扎哈·哈迪德的赞誉"（它）是一本美术馆的教科书"。那时编写这样一本书要花一大笔钱，还得找外国的专业机构帮忙，而我的胆大在于我不相信别人能代替我们业主的头脑与责任，于是乎遵循自力更生的法则自己动手，基于有限的资源和困难的条件，这在我们的同行看来仍属土法上马，实属一件稀罕事。

编出《国家美术馆建筑设计任务书》花了半年时间，克扬打主力，聘请了中央美术学院建筑系程启明教授的研究生刘伟、刘兰婷、付海，北京大学建筑系研究生韩德泉和中国人民大学文学院学生姜山负责专业方面的工作，我做后盾，再加上邵菁菁提供素材，杨济瑜做帮手，基建处长李汉鹏提供了行政保障，所以能集中精力做成这件事。通过这次编写，我们新馆建设团队基本上就训练成熟而搭建起来了。由于饶权、王玮先生对我们这本设计任务书的推介，再加上时任中国美术馆馆长范迪安先生的赏识，给我们增添了出版这本书的信心。我琢磨能把这本工作手册像"哈佛案例"一样介绍给大家，以起到推动这个行业发展的作用。因为在中国各地建美术馆的大气候下，也引来了很多的征询者，更坚定了我们出版此书的信心。在此过程中有赖平面设计师王学军的主持设计，Jacob Dreyer 和美籍华裔青年王莉荔进行英文翻译和审校，使得这本书的面貌更加精致与完善。

书稿沉淀了两年，其间我们经历了无数次与世界著名建筑师的交流和碰撞，也给我们提供了更多的智力支撑，丰满了书的内容。当然也有对此书提出不同意见者，例如我的好友建筑师齐欣，他认为没有必要用这些基础的问题来教育建筑师，因为成熟的建筑师应该熟悉了这些问题。针对这些意见，我还是谨慎认为这本书其实是一本基于国家美术馆建设而提出问题的行动手册，它不是"教训"别人的书，更不可能面面俱到，但对于没有全面思考过在中国兴建美术馆所涉及问题的人们而言，这本书至少是对于相关议题的梳理，一个交流的平台——可能也会导致真的答案。所以我们坚持了出版的初衷、再加上生活·读书·新知三联书店的鼎力支持，这就促成了这本书的最终落地。

对上述我提及的那些参与编写工作的同事，我要在此向你们表示歉意，因编写这本书时严厉地批评过你们，给你们带来的工作压力，现在想来就像黄世仁一般。同时，对那些以各种方式为编写工作出过力而没能提到名字的同事们，我要郑重表示深深的谢意，正是有你们的支持才有了这本书的诞生。

中国国家画院副院长　原中国美术馆副馆长　谢小凡

2014 年 12 月 26 日

Why do we publish
A Field Guide to Chinese Museum Design?

During the ten years from building the Central Academy of Fine Arts
(CAFA) Art Museum to organizing the international bidding for National Art
Museum of China (NAMOC) new venue project, I learned a lot. In 2008, I
wrote a book titled *Museums for Exhibitions*, intended as a handbook for
those who want to build museums; it was well received, and underwent a
second printing. Subsequently, as the organizer of the international bidding
process to select designing proposal for NAMOC's new venue, our first
step was to compile design guidelines that international architects could
understand and use. The field research conducted during the early stage
showed that ideas formed in the past were not well developed. Although
I knew my direction, the lack of thinking paths became a practical issue.
Fortunately, Director Fan Di'an and Professor Zhu Qingsheng highly
recommended the young talent Tang Keyang to me, who thereafter
became the most important team member to tackle problems. Seven
years ago, he had just graduated from Harvard, and became the curator
for "China Pavilion" at 2010 Venice Biennial; he was full of ideas and
extremely competent, and joined our team in compiling this book of
guidelines. Before he joined us, I drafted an outline for the guideline
book; and I subsequently attended a high-profile meeting to discuss the
new national art museum project. From the meeting, I summarized four
requirements: first, it is a Chinese art museum, with prominent Chinese
elements; second, it is a contemporary museum, with features that reflect
current times; third, the museum is practical and functional while being
stately and elegant; fourth, the museum is ecologically sound. To illustrate
the four directive requirements in the design guidelines, that is Dr. Tang
Keyang's task.

The philosopher Zhu Xi of the Southern Song Dynasty advocated the Confucian concept of "studying the world in detail objectively to acquire its principles and standards". In the last few centuries, however, that has not got into the mainstream of Chinese scholarship, which has been dominated by holistic, ambiguous modes of research. However, Dr. Tang always maintains rationalism and objectivity, specifying details and sparing no pains in the development of concrete objectives. His examination of the "museum" and the "Chinese museum" transformed abstract concepts into concrete objectives. He articulated what our art museum needs by specifying details in form, color, space, and circulation patterns, and gave world top architects clear direction and concrete objectives in museum design from the perspectives of museum philosophy, function, management, and construction. This offered an adequate way for us to guide architects to the most suitable designing, instead of the other way around.

Our international bidding competition had two major stages. The first stage called for conceptual schemes, asking architects to come up with creative schemes freely. The second stage narrowed the competitors down to five finalists; in this stage, Architectural Design Guidelines for National Art Museum of China, compiled by Dr. Tang Keyang, was used, and was praised by many renowned figures, including Zaha Hadid, who said that the book was like a textbook for museum design. At that time, writing such a guideline book seemed costly and unprecedented for us, requiring the help of experienced foreign professional institutions. However, I believed that nobody was able to take our minds and responsibilities; instead of outsourcing the job, we decided to work the matters out by ourselves. In the eyes of our colleagues, this was a mission impossible, but was accomplished with limited resources.

It took half a year to compile the book; during this process, Dr. Tang played the major role, and I was the backup. We included graduate students in our team as assistants, including Liu Wei, Liu Lanting, Fu Hai, Han Dequan, and Jiang Shan from the Central Academy of Fine Arts, Peking University, and Renmin University of China. Also, from NAMOC Mr. Yang Jiyu and Miss Shao Jingjing provided the resources needed for the project, while Mr. Li Hanpeng provided administrative support vital to the completion of the project. Through the compilation of this book, we also built up our mature team for the entire National Art Museum project. Thanks to Mr. Rao Quan and Mr. Wang Wei for their appreciation of our guideline book, and to the then NAMOC director Fan Di'an for his trust on us, we have been able to publish this book with confidence. I was hoping to introduce to readers this book, like a Harvard type of case study, to promote the museum design practice. As China's museum fever has led to great curiosity on the topic, we felt that the timing to publish this book was right. Finally, this book would not have been possible without the graphic designing from Mr. Wang Xuejun and the English translation and proofreading from Jacob Dreyer together with Chinese-American Lily Wang.

The book draft sat on the shelf for two years; during this period, we had conversations and encounters with world famous architects, providing us with intellectual stimulation and challenges which enriched the book. Needless to say, contesting opinions emerged; for example, my good friend Mr. Qi Xin dismissed the book as unnecessary, since matured practitioners are already broadly familiar with these issues. With regard to critiques such as his, I still consider this book a guideline book by nature, based on the issues emerging from the NAMOC new venue project, aimed at forming a set of design practice standards. It is not the book's intention "criticizing others" or being encyclopedic. However, for those who have never considered comprehensively the issues at stage in the design of Chinese museums, this book offers a well-woven index of relevant aspects, as well as a platform for further communication, which may result in insightful conclusions. So, we have been pursuing this original goal; with the help of SDX JOINT PUBLISHING COMPANY, this book is finally published.

For the colleagues mentioned above, I would like to apologize for pushing you too hard and putting you under enormous pressure. Finally, for those who have contributed in various ways but are not mentioned here, I would like to express my deep gratitude. Without you, this book would not have been possible.

Xie Xiaofan

Deputy Dean, China National Academy of Painting

Former Deputy Director, National Art Museum of China (2010 – 2018)

2014. 12. 26

方法论
Methodology

○ 我们的讨论将从"此时此地"开始——在 21 世纪的中国。
Our discussion starts from here and now – in China in the 21st century.

○ 什么是美术馆？听听别人的说法。
What is an art museum? Let's listen to others.

○ 每座美术馆都属于它所在的城市。
Each museum belongs to the city where it is located.

○ 美术馆首先是关于看的。
Museums, first, are about ways of seeing…

○ 美术馆也是关于行动的。
… and about ways of doing, too.

○ 在美术馆馆长心目中它是各种功能的集合体。
Museum directors consider combining all potential functions.

○ 对建筑师来说重要的是美术馆如何造出来。
Architects focus on how to build museums.

○ 还得有一些不同的视角……
Perspectives beyond these are important to keep in mind.

○ 所有的讨论最终将会回到实际中去检验。
Discussions will ultimately be evaluated on the basis of practice.

○ 美术馆设计是系统性思维和跨学科合作，是实务更是哲学。
Museum design involves systematic thinking and inter-disciplinary cooperation; it is a practice and a philosophy as well.

注：在英文中，"美术馆"、"艺术博物馆"和"博物馆"这几个概念经常换用。中文则有所不同。本书将根据具体语境选择用词。

Note: In English, "museum", "art museum", and "fine art museum" can sometimes be interchangeable, while in Chinese it is not the case. This book chooses proper wordings according to the context.

第一章

"此时此地"的美术馆
Museums, Here and Now

○ 强调"此时、此地"是为了化繁为简。
We emphasize "here and now" to simplify things.

○ 在中国建立新的美术馆需要满足必要的文化前提。
We have to satisfy necessary cultural conditions in order to
build museums in China.

○ 时间上，美术馆在当地文化传统中（本该）是承上启下的。
In terms of time, museums can (and should) connect local
cultural traditions of the past with the future.

○ 空间上，接地气的美术馆（本该）是当代城市生活的一部分。
In terms of space, locally connected museums can (and
should) become a vital part of contemporary urban life.

○ 可是现在美术馆显然有着不少错误歧异的时空。
Currently, museums which seem lacking in context – spatial
and temporal – abound.

○ 美术馆是"保鲜盒"还是"冷藏室"？
Is a museum a storage box to keep freshness, or a
refrigerator?

○ 在美术馆中展示的是文物，是艺术，还是空间本身？
Does a museum display historic relics, art, or the space itself?

○ 建筑形象和展示的内容之间应该有关系吗？
Is there a relationship between the image of a museum and its
content?

○ 建筑和展示的技术是否和展品彼此呼应？
Can architectural technologies and display technologies be
meaningfully in dialogue with a museum's art objects?

○ 比起西方人，则更有理由问自己这些问题。
Compared with westerners, there are greater reasons for us to
ask these questions.

O

过去时代有美术馆吗?
Were there art museums in the past?

··········

过去时代固然有收藏和展示艺术品的空间，但它们未必是向全社会开放的场所，也没有专业文化机构和制度，支持相关的艺术活动。

··········

In ancient times in China, spaces for art collection and display did exist, but they were not necessarily open to the public; there were no professional cultural organizations and institutions capable of supporting art events.

○ ○

什么样的房子都可以作美术馆吗?
Are all buildings suitable for museums?

○ ○ ○

人们为什么要去美术馆?
Why do people go to museums?

·········

不是任意一座出色的建筑物都可以充任或改建成美术馆。目前大多数优秀的美术馆建筑符合以下三个特征：支持美术品陈列的常规要求，以艺术展览为"主角"，以及在一定程度上引领当代艺术的观念。

·········

尽管对艺术的喜爱古今皆然，欣赏艺术的方式与场所却发生了很大的变化，这种转变是美术馆的概念必须放在特定的时空里看待的原因。

·········

Not just any building can be converted into a museum, regardless of whether it is well-built or not. Most outstanding museum buildings share three features: they are spaces supporting conventional requirements for art display, taking art exhibition as their primary concern, and leading the development of contemporary art concepts to some extent.

·········

Although people have appreciated art from ancient times till the contemporary era, the ways and spaces to appreciate art objects have changed greatly with time, necessitating an examination of the museum concept in spatial and temporal context.

展示空间的中国小史

中国城市的现代化

1840 年鸦片战争之后，中国开始有了西方形态的城市，城市形态的改变急剧地改变了城市生活的内涵。近 200 年后的今天，我们所居住的城市已经和我们祖辈的截然不同了，挑选那些我们最相关的部分来说：

1. 城市构成的变化决定了美术馆是一类前所未有的大型公共空间，混杂于性质不明的半现代半传统街区内；
2. 城市意象的变化决定了美术馆既是"看"的场所也是"看"的对象；
3. 建筑风格的变化决定了西式美术馆建筑物通常是卓然自立的"物体"而非自成一体的"环境"，强调差异而非和谐。

公共空间和展览建筑

美术馆是中国城市中出现的一类新型公共空间。人们普遍倾向于把大多数博物馆和其他展览建筑看作当然的"公共建筑"（就其"外在"的规模和功能而言），但是在建筑史上，现代博物馆的起源却和私人化的建筑类型密切相关（就其"内部"的尺度和经验而言）。

....

A Concise History of Chinese Art Spaces

The Modernization of Chinese Cities

Since the Opium Wars in the 1840s, cities reminiscent of Western style have begun to emerge in China. Urban typologies have changed, rapidly bringing about different styles of urban life. 200 years later our cities have become drastically different from those of our ancestors. Here are some most relevant points:

1 The changed urban fabric has determined the museum's role as an unprecedentedly large public space, located in an undefined blend of modern and traditional building blocks;

2 The changes in its urban role have determined the museum being not only a venue for "seeing", but a subject to "be seen".

3 The changing architectural styles have seen western museum buildings forming usually a set of individual "structures", rather than a holistic "environment", emphasizing individuality rather than harmony.

The Architecture of Public Space and of Exhibitions

The art museum is a new type of public space which has emerged in Chinese cities. People tend to view museums and other display type of buildings naturally as public spaces (due to their "external" scale and function). However, museums historically emerged from private spaces (due to their "internal" scale and experience).

南通博物苑

被公认为中国最早的博物馆，其实是一栋普通的西洋建筑。

Nantong Museum

Nantong Museum is commonly recognized as the first Chinese museum, actually
housed in a plain Western-style building.

南京博物院

由受过西方教育的第一代中国建筑师代表人物梁思成参与设计的具有"民族风格"早期作品。

Nanjing Museum

Nanjing Museum is an early example of the "National Style" of Chinese museums, co-designed by
Liang Sicheng, representing the first generation of Chinese architects with western background.

北京铁路博物馆

这所在北京火车站原址改建的博物馆展现建筑情境与展出对象的统一. 旧车站展出铁路的历史。

Beijing Railway Museum

This museum is located in the original Beijing Railway Station, showing an architectural
context matching its exhibited object – the old station exhibits its own history.

故宫博物院
在原语境中展出原物

The Palace Museum
In this museum, the original objects are exhibited in their original context.

北京展览馆

这座由苏联专家援建的带有西方新古典建筑特点的展览建筑，是具有纪念性的城市地标，除了展览之外，
还是重要的城市公共空间。

Beijing Exhibition Hall

This exhibition building in the Neo-Classical style was designed by Soviet experts. A
monumental urban landmark, it is one of the most important urban public spaces, with a value
beyond exhibitions.

全国农业展览馆

这座和中国美术馆在同一时期建成的"十大建筑"之一的"博览会建筑"具有中国风貌。

National Agricultural Exhibition Center

One of the "Ten Great Buildings" built during the same period as National Art Museum of
China, this is an exhibition building in Chinese style.

南越王墓博物馆
这座基于考古遗址的展览建筑，是另一例"原地展原物"以及"遗址即是博物馆"。

The Museum of the Nanyue King's Tomb
This museum, built on its original archaeological site, is another example of "the original objects on the original site" and "the archaeological site as a museum".

鹿野苑石刻艺术博物馆
新的建筑实践中出现的新型展览建筑，私人收藏，主题藏品，具有戏剧性的展出空间。

Luyeyuan Stone Sculpture Art Museum
This is a new type of museum out of new practice of architecture, with a private collection, a theme on Buddhist art, and a dramatic exhibition space.

苏州博物馆
由著名建筑师贝聿铭设计的具有中国特色的展览建筑。

Suzhou Museum
This is a museum with Chinese characteristics designed by the renowned architect I. M. Pei.

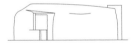

中央美术学院美术馆
这座由矶崎新设计的美术馆是艺术的空间＋空间的艺术。

Central Academy of Fine Arts (CAFA) Art Museum
Designed by Arata Isozaki, this art museum is a space for art as well as a space as art.

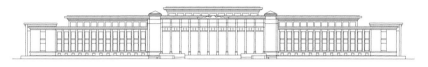

中国国家博物馆

基于原中国革命历史博物馆的扩建，是中国展览建筑增扩建的最有名实例。

National Museum of China

An expansion of the Museum of the Chinese Revolution and History, this is the most
renowned example of museum expansions in China.

中国展览空间 案例介绍
Cases: Chinese Museum Spaces

Cases: World Museum Spaces

南京博物院
Nanjing Museum

鹿野苑石刻艺术博物馆
Luyeyuan Stone Sculpture Art Museum

广东时代美术馆
Guangdong Times Museum

中央美术学院美术馆
CAFA Art Museum

北京 798 艺术区
798 Art District, Beijing

中国美术馆
National Art Museum of China

南京博物院
Nanjing Museum

地点：	南京市玄武区中山东路 321 号
设计者：	徐敬直设计，梁思成修改
建成时间：	1948 年 4 月第一期工程
占地面积：	35000 平方米

Location:	321 Zhong Shan Dong Lu, Xuanwu District, Nanjing
Designer:	Designed by Xu Jingzhi, Modified by Liang Sicheng
Built Time:	April 1948 (Phase I)
Area:	35,000 Square Meters

作为原中央博物院的正殿，这座民国时期的"展览建筑"有着仿辽代建筑的外观，但是它的内部空间和现代功能却和仿古建筑的面貌有着较大的差异。

As the main hall of the original Central Museum, the "exhibition architecture" was built during the Republic of China period with a façade reminiscent of Liao Dynasty architecture, in contrast to its modern interior space and functionality.

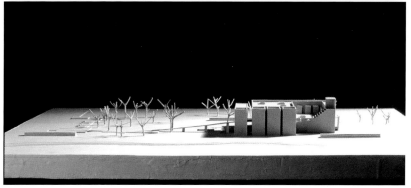

鹿野苑石刻艺术博物馆
Luyeyuan Stone Sculpture Art Museum

地点：	成都郫县新民镇云桥村
设计者：	刘家琨
建成时间：	2002 年 7 月
占地面积：	990 平方米

Location:	Yunqiao Village, Xinmin, Pixian, Chengdu
Designer:	Liu Jiakun
Completion:	July 2002
Area:	990 Square Meters

这座诞生于 21 世纪的私人美术馆在几个方面开创了新的潮流：它基于已有收藏的设计，美术馆有一个"主题"，美术馆和它的郊区环境形成某种对话。

This private art museum, founded in the twenty-first century, started a new trend of art museums in several aspects: it is based on an existing collection, with a specific "theme", and is in dialogue with its suburban environment.

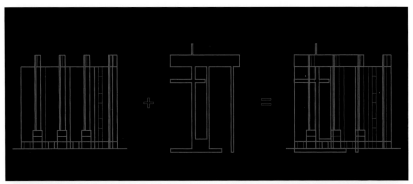

广东时代美术馆
Guangdong Times Museum

地点： 广州市白云大道黄边北路时代玫瑰园三期
设计者： 雷姆·库哈斯 (OMA)
建成时间： 2010 年 10 月
占地面积： 8000 平方米

Location: Times Rose Garden III, Huang Bian Bei Lu, Bai Yun Da Dao, Guangzhou
Designer: Rem Koolhaas (OMA)
Completion: October 2010
Area: 8,000 Square Meters

荷兰建筑师雷姆·库哈斯设定了一个非同寻常的美术馆概念：把美术馆建在一个真正的生活场景中。这座城市美术馆实际占据一栋普通高层住宅的若干层，彼此由专门电梯连接。虽然美术馆的使用一定程度上独立于居住空间之外（主要展厅位于顶层），但这种把美术馆真正"融于生活"的做法在世界范围内也是极为罕见的。

Dutch architect Rem Koolhaas envisioned an unusual concept: inserting an art museum into a real-life scenario. This urban art museum occupies several levels of an ordinary residential building, interconnected by exclusive elevators. Although the use of the art museum is separate from the residential space to some extent (the main exhibition space is on the top floor), the idea of "blending" an art museum into daily life is unique even globally.

中央美术学院美术馆
CAFA Art Museum

地点：	北京市朝阳区花家地南街8号，中央美术学院内
设计者：	矶崎新
建成时间：	2008年3月
占地面积：	14777平方米

Location:	CAFA Campus, 8 Hua Jia Di Nan Jie, Chaoyang District, Beijing
Designer:	Arata Isozaki
Completion:	March 2008
Area:	14,777 Square Meters

相对于传统的美术学院"画廊"，中央美术学院美术馆的建筑设计和内部空间自身就是某种"艺术"。

In contrast with the traditional galleries of art academies, Central Academy of Fine Arts (CAFA) Art Museum's architecture design and interior space are themselves "artworks" of certain type.

北京 798 艺术区
798 Art District, Beijing

地点:　　　北京朝阳区大山子
设计者:　　东德建筑师初建，中国建筑
　　　　　　师改建
建成时间:　1957 年至今
占地面积:　600000 平方米

Location:　　Dashanzi, Chaoyang District,
　　　　　　　Beijing
Designer:　　Designed by GDR architects,
　　　　　　　renovated by Chinese architects
Completion:　1957-present
Area:　　　　600,000 Square Meters

不是一座独立的美术馆，而是一个美术馆的聚集地。北京798 艺术区是在 1950 年代由当时的东德援建的工厂厂区基础上形成的。它既涉及旧城市街区再开发的"缙绅化"问题，又把美术馆设计的单纯建筑议题扩大到城市规划和城市设计领域。

Not only one art museum, 798 Art District is a cluster of art museums. Converted from a former factory complex built during the 1950s by the then-East German architects, it has offered solutions to issues of "gentrification" in the redevelopment of old urban blocks, as well as issues of extending museum design to urban planning and urban design frameworks.

中国美术馆
NAMOC

中国美术馆设计介绍
Introduction to the
Design of the National
Art Museum of China

地点：	北京市东城区五四大街 1 号
设计者：	戴念慈
建成时间：	1963 年 3 月
占地面积：	17051 平方米

Location:	1 Wu Si Da Jie, Dongcheng District, Beijing
Designer:	Dai Nianci
Completion:	March 1963
Area:	17,051 Square Meters

已有约六十年历史的中国美术馆在 2010 年代考虑建设一座新馆，对于美术馆建筑也面临着回顾历史和更新认识的机遇。

The National Art Museum of China (NAMOC) has a history of about 60 years and was considering building a new museum in 2010s. It was facing an opportunity to review the history and refresh notions of art museum architecture.

1. 城市建筑尺度的跃进

在传统中国城市中，单体建筑之间的差别并不显著，而且这种差别从院落外是看不见的，高等级的建筑和普通建筑在体量上相差无几，基于此相似性，建筑单元靠平面上的互相组合形成更大的单元。

1963 年建成的中国美术馆是这一区域最引人注目的建筑，形成了以它为中心的城市空间，这座建筑显著地大于它附近的民居，它的建筑面积超过周围院落面积的总和。建筑体量的急剧扩大和对于公共性的要求，使得中国美术馆的体量和空间组织方式显著区别于传统城市中的"单体建筑"。

A Great Leap Forward in Building Scale

In a traditional Chinese city, individual buildings are relatively similar to each other, with few discernable differences externally. Expensive constructions and ordinary buildings have little variance in scale and volume. Because of their similarities, individual building units could therefore be integrated or fit together into larger combinations.

Built in 1963, the current National Art Museum of China (NAMOC) is one of the most prominent buildings in its area, creating a unique focal point of urban space. This building is demonstrably larger than its neighboring residential buildings, with its construction area far exceeding the sum of the latter. The significant increase in building volume and the needs for public space made NAMOC manifestly different from "individual buildings" in traditional cities in terms of volume and space organization.

1　中国国家美术馆城市区位
Location of NAMOC

2　中国美术馆东南 – 北，东 – 西剖面，戴念慈绘制
Section, SE-N, E-W, by Dai Nianci

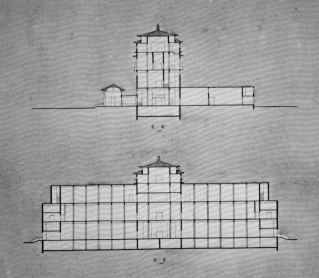

剖面图
1:200

剖面图

中国美术馆圆厅（改造前）
NAMOC Rotunda (before renovation)

2. 包罗万象的"中国美术"

中国美术馆自从建馆 60 余年以来，至今共有十余万件藏品，其中油画 2625 余件，中国画近 8298 件，版画 6000 余件，雕塑 1000 余件，水彩、摄影、陶瓷、漆画、工艺品等其他艺术作品 6000 余件，各类民间艺术品 60000 余件。藏品中最大体积尺度为 4m×4m×3m，最大平面尺度为 9m×3.5m，最大重量为 4.5 吨。

A Big Picture of "Chinese Art"

With a history of about 60 years, NAMOC has built an art collection of more than 100,000 pieces, including 2,625 oil paintings, 8,298 Chinese paintings, about 6,000 prints, about1,000 sculptures, over 6,000 other fine art works (including watercolor, photography, ceramics, lacquer paintings, crafts and so on), and about 60,000 folk art works. The maximum volume is 4×4×3m. The maximum size of two-dimensional works is 9×3.5m. The maximum weight is 4,500 tons.

中国美术馆建成不久后的主立面
Main façade of NAMOC shortly after its completion

中国美术馆设计渲染图，戴念慈绘制
NAMOC rendering drawing, by Dai Nianci

3. 古典式样，现代功能

1959 年前后，为庆祝中华人民共和国成立十周年，前后建成了号称
"十大建筑"的十座首都地标式建筑物。中国美术馆由戴念慈设计，
1963 年建成，建筑借鉴了敦煌的"九层楼"意象，为"民族风格"
的现代建筑的代表作品。

"民族风格"式样的"十大建筑"同时具有中国和西方古典建筑的某
些特点。它保持了中国古代建筑的比例特征和造型特点。

Classical Style for Modern Functions

To commemorate the 10th anniversary of the founding of the
People's Republic of China, ten capital landmarks, known as the
"Ten Great Buildings", were erected around 1959. The National
Art Museum, designed by Dai Nianci, was completed in 1963.
Channeling the Dunhuang architectural image of the "Nine-Level
Pavilion", it became a representative piece of the "National Style" of
modem architecture.

The "Ten Great Buildings" in "National Style" have retained traditional
Chinese architectural features in terms of proportion and shape,
while also echoing certain classical features in Western architecture.

1   "盛世和光"大展中，中国美术馆的立
    面被打扮成敦煌石窟的形象，体现它和
    原型的联系。

    NAMOC cladded in a Dunhuang
    mock-up in the "Glory of Golden
    Age" exhibition, connecting itself
    with its architectural prototype

2   敦煌莫高窟"九层楼"立面
    The "Nine-Level Pavilion" façade of
    Dunhuang Caves

3   中国美术馆建筑模型
    The model of NAMOC building

第二章

什么是美术馆

What is a museum?

○ 什么是美术馆?
What is an art museum?

○ 什么是美术，什么是艺术?
What is fine art, and what is art?

○ 如何区分博物馆、美术馆、画廊、会展空间或是拍卖场?
What is the difference between a museum, an art museum, a gallery, an exposition space, and an art auction space?

○ 美术馆不仅有展场，还要考虑展品、展览和展事。
An art museum considers not only a display space but also collections, exhibitions and events.

○ 不同的时空中，即使是西方的展览空间也是面貌不一的。
The Western practice of museums has evolved with time and space.

○ 博物馆和美术馆的区别在于看待展品的方式。
A museum differs from an art museum in the way of how it considers art objects.

○ 当代美术馆既是展示的空间也是交流的场所。
Contemporary art museums are spaces not only for exhibition, but also for communication.

○ 画廊里不止是画。
Galleries are not just for paintings.

○ 某种意义上，是艺术造就了美术馆建筑。
In a sense, it is the art that makes the art museum.

○ 美术馆的历史是一部依然在发生变化的社会史。
The history of art museums is a social history, one which continues to evolve.

展品
186 中国艺术传统的影响 The Influence of Chinese Art Tradition

Art Objects

·········

展品大多来自于藏品，两者既有区别又有联系。在中国，美术馆的展览对象可以千差万别。随着艺术作品定义的不断丰富，它的品类已从以"国油版雕"为代表的美术作品拓展至各种综合性的视觉艺术门类，如装置、影像、现当代设计等。展示的对象不一定可以收藏，有可能是"临时"的。

·········

Art objects are mostly from collections; they are related but not the same. In China, art objects included in a show are diverse. With the enrichment of the definition of art objects, the categories have expanded from the traditional "Chinese ink painting, oil painting, engravings, and sculpture" to a variety of visual art categories, such as installations, video art, contemporary design, and so on. The art objects of an exhibition are not necessarily collectible. They may be temporary or ephemeral in nature.

不同展品尺寸的对比
Artworks of Different Sizes

泉（现成品）/ 36×48×61cm
Spring (ready-made object)

兰亭集序（书法）/ 70×30cm
Preface to Lanting Gathering (calligraphy)

蒙娜丽莎（油画）/ 53×77cm
Mona Lisa (oil painting)

唐三彩骆驼（陶器）/ H 87cm
Tang Three-Colored Camel (pottery)

后母戊大方鼎（青铜器）/ H 133cm
Houmuwu Ding for Ritual Use (bronze)

断臂维纳斯（雕塑）/ H 203cm
Venus de Milo (sculpture)

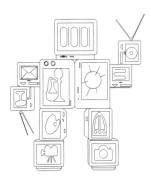

Globe（装置）/ 165×153×13cm
Globe (installation)

○ ○

展览
Exhibitions

....

"展览"是美术馆空间使用的基本形式，但美术展览很难有标准定义的
内容和形式。中国所说的"现代"或"当代"美术馆展出的对象在时
间段上主要为 20 世纪以来的艺术，但同时也可能包括以古典艺术为灵
感或内容的展览；一般来说，美术馆注重专业艺术家和学院艺术家的
作品，同时，各级公立美术馆也经常举办大量由中国各企事业单位和
行业协会牵头组织的展览。这些作品的内涵不一定等同于西方美术馆
认定的"现代艺术"。展览方式由古典到近代，展览构思由封闭到开放，
展览环境由私人到公共。

....

Holding exhibitions is the essential use of the museum. It's difficult
to precisely define the content and formation of exhibitions; in China,
modern and contemporary art museums have held exhibitions of works
made since the 20th century as well as works inspired by classical art
and of classical art itself. Generally speaking, art museums' focus is on
the work of professional artists and artists from academia; meanwhile,
public museums at all levels hold exhibitions initialized by Chinese
enterprises and professional organizations. The meanings of these
works are not necessarily the same as those defined by Western notions
of modern art. The curatorial strategies of exhibitions range from the
classical to the modern, the conception from conservative to open, and
the display environment from private to public.

不同展览环境的简单示意
Diagrams of Exhibition Settings

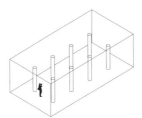

古典展示（柱廊）
Classical Display (Colonnade)

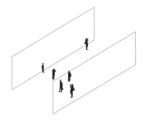

近现代展示（画廊）
Modern Display (Galleries)

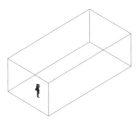

自然光照明（白盒子）
Naturally-lit (White Box)

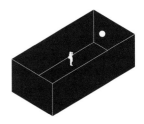

人工照明（黑盒子）
Artificially-lit (Dark Box)

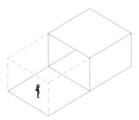

室外 / 半室外空间（灰盒子）
Uncovered/Partially-covered Space (Grey Box)

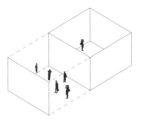

公共 vs 私人空间（嘈杂 / 安静）
Public vs. Private Space (Noisy/Quiet)

··········

美术馆是围绕着艺术展览的公共活动中心和专业研究中心。典型活动如下：

- 开幕式
- 展览研讨
- 小规模艺术展首展观摩
- 公共教育
- 配套放映和演出
- 艺术家在馆创作

··········

Art museums are centers of public activities and professional research based on art exhibitions, with typical events as follows:

- art exhibitions openings
- exhibition conference
- small-scale art premieres
- public education
- screening and performance
- artists-in-residence

公共教育
Education
国博讲堂——历史与艺术的体验系列活动（2012年度暑期专题）。
National Museum lecture series in 2012 summer

专业培训
Training
2012年6月2日，由文化部艺术司主办，全国美术馆专业委员会、浙江美术馆、中国美术学院承办的2012年全国美术馆高级管理人员（馆长）培训班（第一期）在浙江美术馆举办。
Training Session for museum directors at Zhejiang Art Museum

学术研究
Research
2012年6月11日中国传统壁画保护、修复、绘制理论研讨会在中央美术学院举行。
Symposium on Chinese traditional mural protection at CAFA

办公
Office
2012年6月26日下午，日本东京艺术大学美术馆馆长关出先生等一行6人受邀对中央美院美术馆进行了访问。
Visit to CAFA Art Museum by Japanese museum directors

展　览

EXHIBITION

商业
Shopping
尤伦斯当代艺术中心商店
Shop at UCCA Center for Contemporary Art

餐饮
Catering
今日美术馆咖啡厅
Cafe at Today Art Museum

作业
Working
泰特美术馆与中国美术馆合作，将透纳（J.M.W.Turner）的80件绘画作品送到中国进行展览。
Tate Museum's cooperation with NAMOC on J. M. Turner's paintings

会议/讲座
Conference/ Lecture
2012年5月11日杉本博司在中央美术学院举办讲座。
Lecture by artist Hiroshi Sujimoto at CAFA in 2012

观演
Theatre
2012年7—8月，广东时代美术馆与香港歌德学院共同合作，为广州观众带来10部赫尔佐格的纪录片，并特邀香港电影及纪录片教育工作者蔡甘铨先生出席映后交流会。
Documentary screening on Herzog at Guangdong Times Museum in 2012

注：本页内容为简译。
Note: Translation on this page is simplified for reference.

世界美术馆小史

没有收藏就没有美术馆，但是"美术"或"艺术"到底是什么，不同时期的人们有不同的看法。现代意义上的美术馆起源于文艺复兴时期的私人收藏，阿拉贡的阿方索家族、佛罗伦萨的美第奇家族收藏都是其中著名的例子。他们的兴趣包括希腊和罗马时期的文物、名人肖像等。严格说来它们并不都是现代意义的"艺术作品"，这导致了这些朴茂的"美术馆"空间上的含混性——不像现代人司空见惯的"白盒子"展厅。在那时的"美术馆"中，错落地放置在柱廊间的雕像，或是应赞助人请求特别创作的壁画，它们到底是家具性的室内装饰品，或是一种空间构成的必要手段，还是独立的艺术创作，在"住宅"的情境之中我们已经很难区分了。美术馆起源中蕴含的这些矛盾也延续到今日的美术馆设计中。

但是人文主义的艺术精神恰恰来源于这种矛盾的空间。这种空间的混合体被恰如其分地称作"私人的万神庙"：一方面，艺术品是人类精神的催化剂，这种认识不可能来自于君主和封建贵族，而只能在独立的"个人"基础上萌发，在这个意义上发展出的鉴赏是亲密的活动；另一方面，文艺复兴时期人们对于"殿堂"的崇敬感延续了古典建筑的传统，当代美术馆甚至也延续了这种宗教背景下的威仪。

美术馆的功能是什么？它既包括展现、喻示、（公共）交流，也包括学习和沉思，如果说展现和谕示是自上而下的宣教，学习和沉思的功能却是依赖于精神上的自觉。介于这两对功能间的公共交流功能是注定有别于娱乐的，它只能是人际尺度和纪念性尺度的折中和平衡，而无法仅仅是其中一种。

....

A Concise History of World Museums

Art collections make an art museum. But what is fine art, or art, really? People at various times hold different opinions. The modern museum originates from private collections during the Renaissance. The Alfonso family of Aragon and the Medici family in Florence were among the most notable private collectors. Their interest was all-encompassing, ranging from Greek and Roman antiques to celebrity portraits – not all of these were "artworks" in the modern sense, giving an undefined character to these rudimentary premodern museums. Unlike the "white box" gallery of today, randomly standing between colonnades in these spaces were sculptures, as well as murals commissioned onsite. Were they truly independent art creations, or merely a furniture-type interior decoration, or a necessary form of spatial composition? It's hard to distinguish, as they all were in the "residence" context. The paradoxes in the origin of art museums are also seen in contemporary museum design.

The spirit of the arts in humanism comes precisely from this paradoxical space of mixed use, called properly "the private Pantheon". On the one hand, artwork is the catalyst for human spirit – this perspective comes not from the nobles or feudal lords, but only from independent individuals; art appreciation in this regard is intimate and personal. On the other hand, the Renaissance age has sustained the classical architecture tradition via awe and respect for "temples"; today's art museums have inherited this kind of ritualistic relationship between viewers and space engendered in religious context.

What are the functions of an art museum? They include exhibition, pedagogy, public communication, learning, and meditation. If exhibition and pedagogy functions mean top-down "public education", learning and meditation functions rely on personal bottom-up initiatives. The communication function is not recreational at all. It has to be a compromise and a balance between the public and the personal perspectives, instead of being one of them.

第一波冲击
文艺复兴以前

希腊、罗马时代至中世纪的公共建筑常常有精美的艺术品陈列，但这些艺术品并不是现代美术馆中的艺术。无论如何，这些建筑和艺术品的关系深刻地影响了现代意义上的美术馆的起源，它们是西方建筑类型学中所说的美术馆空间"原型"。

The First Wave
Pre-Renaissance

From the Greek and Roman periods and until the medieval age, public buildings often were used to display refined artworks, but these were not necessarily equivalent to the display of modern art in contemporary museums. Nevertheless, the relationship between the built environment and the exhibition of artworks formed during this time continues to make a strong impact on contemporary museums, casting the typological "prototypes" of Western museums.

第二波冲击
文艺复兴

文艺复兴是一般公认的现代美术馆的起点。在这一时期的私人美术馆沿用了其他功能建筑的物理样式，但是美术与艺术品成为美术馆空间的焦点时，这些建筑原有的属性也正在发生变化。

The Second Wave
Renaissance

The Renaissance is commonly recognized as the starting point of modern museums. In this period, private museums inherited building typologies with alternative functions; but when artworks became the focus of museum spaces, the original properties of these spaces changed along with them.

第三波冲击
文艺复兴至 20 世纪中叶

The Third Wave
Renaissance to Mid-20th Century

随着现代资本主义国家的兴起，美术馆成了大城市的文化建筑"标配"。艺术展览建筑的物理样式和功能程序趋于成熟。

With the rise of the modern capitalist states, art museums became a standard cultural element of the western metropolis, with a physical type and an architectural program gradually developed.

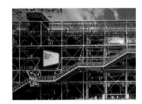

第四波冲击
20 世纪后半叶至今

20 世纪初发展起来的现代艺术已广泛获得社会认同。随着艺术定义的变化，美术馆建筑类型趋向多样化，艺术空间创新的同时也促进了艺术自身的变革。

The Fourth Wave
The 2nd Half of the 20th Century to Present

Modern Art, developed since the early 20th century, has been widely accepted in Western society. As the definition of art changes, museum types diversify; architectural innovations of art spaces simultaneously promote the revolution of art itself.

世界美术馆分类介绍
Cases: World Museum Spaces

希腊帕特农神庙和罗马万神庙
Parthenon and Pantheon

雅典和罗马
Athens and Rome

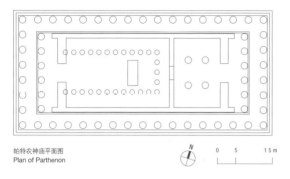

帕特农神庙平面图
Plan of Parthenon

N

0   5   1 5 m

帕特农神庙是多立克柱式与爱奥尼亚柱式的完美结合。古典"柱式"形成的内廊确立了人们顺序体验空间的线性模式，是后世美术馆"画廊"单元的原型。而罗马万神庙代表的富有纪念性的整一空间，则反复出现在类似于"圆厅"这样的美术馆单元中。

The Parthenon is a perfect combination of the Doric and Ionian Orders; Classical orders formed the colonnades, establishing a basic linear pattern for people to experience space via order. It became the prototype for the "galleries" of the museums which followed it. A unified and monumental space, best represented by the Roman Pantheon, is often found in the rotunda typology of modern art museums.

在古典时代，纪念性的空间本身就是展示的对象。

In the Classical period, monumental spaces themselves were exhibits to display.

卫城中的艺术品已移至它处，卫城变成了一座没有展厅的博物馆。
Artworks were removed from Acropolis, which now becomes a museum without galleries.

乌菲齐美术馆
The Uffizi Gallery

意大利佛罗伦萨
Florence, Italy

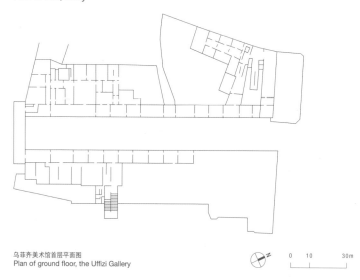

乌菲齐美术馆首层平面图
Plan of ground floor, the Uffizi Gallery

0   10   30m

乌菲齐美术馆始建于 1560 年，美术馆的主体部分是三个几乎平行的条状空间，并联着两侧的房厢次第展开，在美术馆中既有古典时期常见的柱间展示，也出现了当时罕见的作为艺术品背景的展墙。

Founded in 1560, the museum is mainly composed of three almost parallel hallways. Each hallway has attached rooms interconnected. This museum has display spaces between the colonnades, as commonly seen during the Classical period; walls are also used for displaying, a practice not typical then.

在文艺复兴时期的展览空间中，艺术品开始占据中心地位，展示环境逐渐变成艺术品的背景或者"画框"。

In the exhibition spaces of the Renaissance period, artworks started to gain spotlight. The environment in which the works were displayed gradually became the background or frame for the works.

乌菲齐美术馆内景
Interior of the Uffizi Gallery

大英博物馆
The British Museum

英国伦敦
London, UK

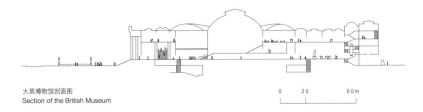

大英博物馆剖面图
Section of the British Museum

0   20   60 m

大英博物馆始建于 1753 年，其后历经数次重新设计和扩建。它是世界上历史最悠久、规模最宏伟的综合性博物馆之一。博物馆主体采用了希腊复兴式的建筑风格，仿效了古希腊的神庙，同时又采用了 19 世纪 20 年代的当时先进的铸铁装置。英国著名建筑师诺曼·福斯特在博物馆中植入的一个圆形中庭，现在用作博物馆的图书馆的一部分。

Founded in 1753, the British Museum has been redesigned and expanded several times. It is one of the most historical, one of the largest, as well as one of the most comprehensive museums in the world. The main venue of the museum has adopted Greek Revival architectural style, mimicking ancient Greek temples. The museum has also employed the then-advanced cast-iron fixtures of the 1820s. The renowned British architect Norman Foster has created a rotunda atrium in the museum, which is used as part of the museum's library.

在西方资本主义国家兴起的过程中，美术馆逐渐成为文明政府和帝国强权的象征，这一时期的展览空间和展示对象往往具有不容轻慢的纪念性特征。它们在继承古典时代和文艺复兴时期此类空间特征的同时，又赋予艺术品有史以来至高无上的地位。

With the rise of the western capitalist states, art museums gradually became the symbol of civilized government and imperial power. Art spaces and exhibits in this period usually carry a monumentality, one which is not easily ignored. While inheriting the characteristics of such spaces from the Classical and Renaissance periods, museums in this period endowed art objects with a status much higher than before.

大英博物馆入口
Entrance to the British Museum

在希腊化时期就有博物馆这样的说法，但是展示的对象、目的和方式都和今天大不相同。

The wording of the museum comes all the way from the Hellenistic period; but the practices with regards to the objects being displayed, the purpose of such display, and the techniques of display employed at that time are quite different from today.

大英博物馆内景
Interior of the British Museum

大英博物馆鸟瞰
Overlook of the British Museum

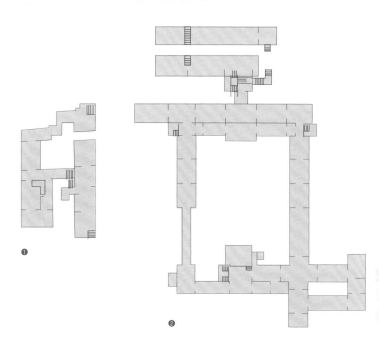

0

0

the British Museum

the British Library

1　大英博物馆三层平面图
　　Plan of Floor 2, the British
　　Museum

2　大英博物馆二层平面图
　　Plan of Floor 1, the British
　　Museum

3　大英博物馆首层平面图
　　Plan of Ground Floor, the
　　British Museum

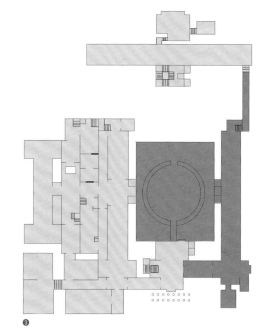

0

纳尔逊 – 阿特金斯艺术博物馆新馆
The Nelson-Atkins Museum of Art (New Building)

美国堪萨斯城
Kansas City, USA

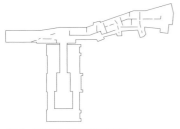

纳尔逊 – 阿特金斯艺术博物馆地下层平面图
Plan of Underground Level, Nelson-Atkins Museum of
Art

0   25   75 m

纳尔逊 – 阿特金斯艺术博物馆始建于 1930 年，1933 年首次对外开放。由于当时正处于经济大萧条时期，全球艺术品市场涌现出大量低价倾销的珍贵艺术品，而收购者很少，该馆因此在很短的时间内收购了大量的艺术品，奠定了在艺术博物馆中的地位。该馆在 21 世纪初聘请建筑师斯蒂文·霍尔设计新馆。霍尔的新馆是若干个互相连缀的半透明发光盒子，狭长的盒子不仅在外观上迥异于原博物馆，也改变了展出环境和观看的原有关系。

Launched in 1930 and opened in 1933, the Nelson-Atkins Museum of Art was born during the Great Depression. Valuable art objects surfaced into the market at much lower prices than before globally, with few buyers. Therefore, the museum purchased a large quantity of artworks in a short period, making it an outstanding American art museum. At the beginning of the 21st century, the museum employed the architect Steven Holl to design a new building. Holl's new building is made of a few semi-translucent boxes in long and narrow forms, interconnected and lit from within, different from the original building not only in appearance but also in the relationship between the display context and the way of viewing.

一座美术馆的建筑风格和陈列方式彼此相关，它们呼应于不同的时代，而且可能因时间的流逝而产生新的需求。

The art museum's architectural style and display technique correlate; they correspond to the needs of different periods and will possibly need changes with the passage of time.

纳尔逊－阿特金斯艺术博物馆，新馆和老馆对比
Contrast of the new building and the old building, the Nelson-Atkins Museum of Art

....

纽约现代美术馆
The Museum of Modern Art (MoMA), New York

美国纽约
New York, USA

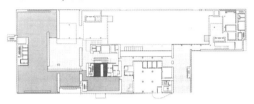

现代美术馆标准层平面图
Plan of typical floor, MoMA

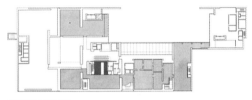

现代美术馆首层平面图
Plan of ground floor, MoMA

纽约现代美术馆位于曼哈顿第 53 街、54 街之间街区密布的摩天楼群中。由洛克菲勒家族在 20 世纪上半叶一手缔造，这座美术馆致力于展览和推广现代艺术。在世纪之交经历了一次主要的改建后，这座美术馆和城市的关系更加密切，现代艺术在其中获得了类似于古典艺术在卢浮宫中的地位。

MoMA is situated in the weblike skyscraper region between 53rd and 54th street in Manhattan. Created and promoted directly by the Rockefeller family during the first half of the 20th century, the art museum devotes itself to the exhibition and promotion of modern art. After one recent major renovation at the turn of the century, the art museum has become closer, in relationship, to the city. In this art museum, modern art has acquired a status equivalent to Classical art in the Louvre.

特定的文明型态造就了特定的城市生活，这种生活既是现代艺术的土壤，也是它的物理容器。

Specific civilization patterns make specific type of urban life, which serves as both the soil and the physical container of modern art.

纽约现代美术馆街景
A street view of MoMA, New York

纽约古根海姆美术馆
Guggenheim Museum, New York

美国纽约
New York, USA

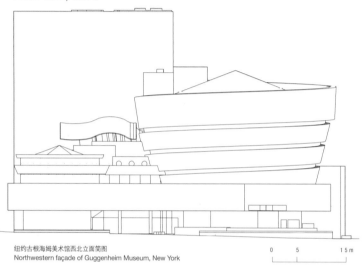

纽约古根海姆美术馆西北立面简图
Northwestern façade of Guggenheim Museum, New York

0   5   1 5 m

纽约古根海姆美术馆是现代主义建筑大师弗兰克·劳埃德·赖特的杰作。
1947 年进行设计，1959 年建成。美术馆独特的连续螺旋型流线创造了
一种与众不同的美术馆空间类型，在其中空间和展览的对象一样具有艺
术价值。

The Guggenheim Museum in New York is the masterpiece of the
renowned modernist architect Frank Lloyd Wright. Designed in 1947
and completed in 1959, the continuous spiral circulation route of the
art museum creates a distinctive space type in art museums, where the
space is as artistically valuable as the artworks displayed there.

美术馆既适应于已有的艺术展览类型，又时常创造出新的展示样式和体验，从而改变艺术自身。

The art museum not only adapts itself to existing art exhibition patterns, but also creates new exhibiting formats and experiences, thus changing the art itself.

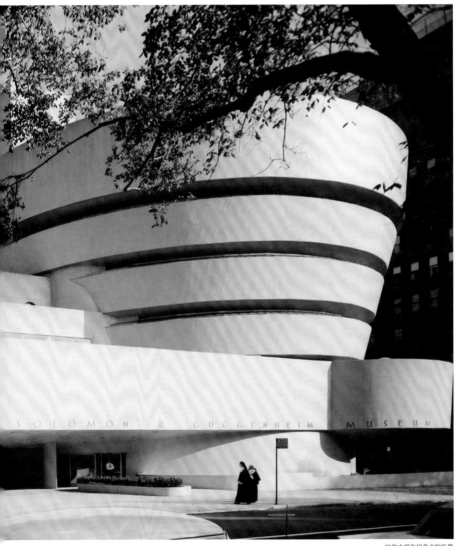

纽约古根海姆美术馆街景
A street view of Guggenheim Museum, New York

乔治·蓬皮杜国家艺术文化中心

The Centre National d'Art et de Culture Georges Pompidou

法国巴黎
Paris, France

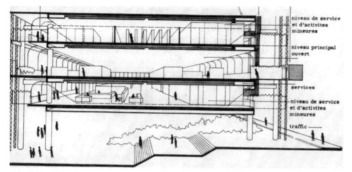

乔治·蓬皮杜国家艺术文化中心剖面透视图
Section and Perspective of Le Centre Pompidou

乔治·蓬皮杜国家艺术文化中心位于巴黎著名的拉丁区北侧、塞纳河右岸，是应法国前总统蓬皮杜的创议于 1977 年建成开放的。建筑的支架由两排间距为 48 米的钢管柱构成，楼梯及所有设备完全暴露。东立面的管道和西立面的走廊均为有机玻璃圆形长罩所覆盖，留出不加掩饰和灵活使用的内部。大厦内部没有传统博物馆那样分割的形成展厅，而是构成若干个整体性的展示空间，并同时设有现代艺术博物馆、图书馆和工业设计中心。

On the right bank of the Seine, north to the famous Latin Quarter, the Centre National d'Art et de Culture Georges Pompidou (Le Centre Pompidou) was proposed by former French president Georges Pompidou and opened in 1977. The building is comprised of two sets of steel pipes with 48 meters of space in between, with staircases and mechanics exposed to the outside. The tubes in the eastern façade and the corridors in the western façade are enclosed by fiberglass tubes, leaving the interior space unembellished and flexible to use. The interior of the building is not divided into exhibition rooms, as those in a traditional museum; rather it forms several large unitary exhibition spaces. The building also houses a contemporary art museum, libraries, and a center of industrial design.

随着建筑创作手法的多样化，美术馆的自身形象和专业功能可以互相统一也可以彼此脱离，或者形成创造性的组合。

With the diversification of architectural approaches, the image of the museum and its professional function can be unified, or detached from each other, or forming a creative combination.

乔治·蓬皮杜国家艺术文化中心外景
Exterior of Le Centre Pompidou

....

盖蒂中心
The Getty Center

美国洛杉矶
Los Angeles, USA

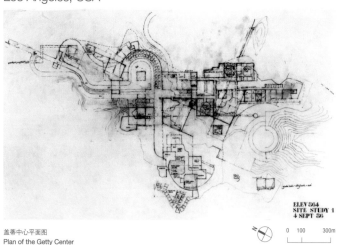

ELEV 864
SITE STUDY 1
4 SEPT 86

0   100   300m

盖蒂中心平面图
Plan of the Getty Center

盖蒂中心位于洛杉矶的布伦特伍德社区山顶。来访者通过轨道火车或驾车到达山顶。1997 年落成开放的中心及其博物馆可以俯瞰洛杉矶全景，它创造了一个集展示、收藏、研究于一体的艺术中心群落。中心的设计者是"纽约五人组"之一的理查德·迈耶，它以明快的色调和现代主义的简洁风格带来展示古典和当代艺术品的不同环境。来自蒂沃里的石灰华大理石强调了创始人对于意大利传统的爱好。

The Getty Center is located on the hills of Brentwood in Los Angeles. Visitors visit the museum by tram or by car. The center and its museum opened in 1997, overlooking the Los Angeles area. They set an example of an art center that integrates functions of exhibition, collection and research. The designer is Richard Meier of the New York Five. The light color and concise Modernist style bring about a unique environment to display classical and contemporary artworks. The travertine from Tivoli is a particular favorite of the founder for its engagement with Italian traditions.

美术馆不见得是一个单一的建筑物，它也可能是由一组建筑物形成的自成一体的人工环境。

The museum is not necessarily a single building; it could be a self-contained man-made environment, composed of a series of buildings.

盖蒂中心鸟瞰
An aerial view of the Getty Center

泰特现代美术馆
Tate Modern

英国伦敦
London, UK

泰特新馆北立面外景
An exterior view of Tate Modern's north façade

泰特美术馆新馆的前身是一个无窗的发电厂，有一座几乎一百米高的巨大烟囱，通过建筑师赫尔佐格和德梅隆的改造，这个原已废弃的砖石建筑物在 21 世纪成了一处世界艺术家们表演的舞台，大小空间结合。其中的巨型"涡轮大厅"常常用于展出一些不同寻常的作品，比如奥拉弗尔·埃利阿松的"气候"。

The new Tate Modern was converted from a windowless power plant with an overwhelming 100-meter-high chimney. With the redesign of Herzog & de Meuron, the abandoned industrial structure became a stage in the 21st century for the world's top artists, with a selection of big and small spaces. Among them is the giant Turbine Hall, often used for uncustomary works, such as Olafur Eliasson's Climate.

美术馆还有可能是由既有的建筑物改造而来。由旧空间改造的美术馆建筑类型，将其原有实用功能进行改造，以满足新时代的文化需求，这是后工业时代老城市"缙绅化"再开发的常见手段。

Museums can be converted from existing structures, in which case architects convert spaces from old buildings, and redevelop original functions to satisfy the cultural needs of a new age. This is a common practice of gentrification of old cities in the post-industrial age.

泰特现代美术馆内景
An interior view of Tate Modern

伦敦水晶宫
The Crystal Palace, London

英国伦敦
London, UK

伦敦水晶宫外景
A view of the Crystal Palace at the 1851 Expo

用铸铁、锻铁和玻璃建成的伦敦水晶宫是1851年伦敦博览会的主场地，开创了近现代博览会建筑的先河。其外观为长方形，呈三级台阶的形状，中央建有一个带筒形拱顶的耳堂，在内部形成了一个巨型的交叉通道，中间种有参天的老榆树。水晶宫不仅仅是一项材料与技术的革新成果，从美学上来看，它也是承前启后的巨大创新。工业化模数（7.2米）的建立，暴露细节的构建、曲线形屋顶和快速装配方式，都给人留下深刻印象。水晶宫代表了工业革命以来用钢铁与玻璃建造的桥梁、温室、厂房和火车库房等建筑传统的巅峰，也是美术馆建筑通往现代建筑之路上迈出的革命性一步。

Made of cast iron, wrought iron, and plate glass, the Crystal Palace was the primary venue for the 1851 London Expo, a prelude to modern expo buildings. Rectangular in shape, it cascaded in three levels. The central aisle, with a curved arc dome as roof, formed a giant tunnel internally. Large elm trees grew in the center. The Crystal Palace is not only innovative in technology and building materials, but also aesthetically extraordinary in linking the past with the future – its introduction of industrialized modules of 7.2 meters, exposed construction details, curved roofs, and speedy installation are highly impressive. The Crystal Palace represents the summit of architecture tradition since the industrial revolution, when glass and steel were used in bridges, conservatories, factories and warehouses. It is also a revolutionary leap in the construction of art museums and the modernization thereof.

"博览会建筑"和当代美术馆有着极近的亲缘关系。

Exposition buildings have a close relationship with the
contemporary museum.

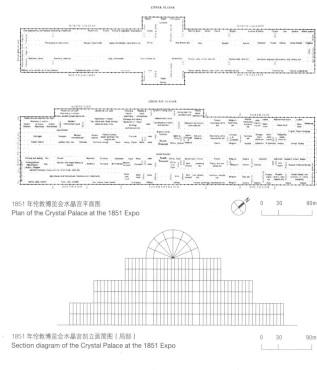

1851 年伦敦博览会水晶宫平面图
Plan of the Crystal Palace at the 1851 Expo

0   30   90m

1851 年伦敦博览会水晶宫剖立面简图（局部）
Section diagram of the Crystal Palace at the 1851 Expo

0   30   90m

1851 年伦敦博览会水晶宫剖立面透视图
Section perspective of the Crystal Palace at the 1851 Expo

蛇形画廊展亭 2010
The Serpentine Gallery Pavilion, 2010

英国伦敦
London, UK

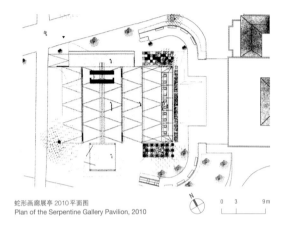

蛇形画廊展亭 2010 平面图
Plan of the Serpentine Gallery Pavilion, 2010

0 3 9m

从 2000 年开始，位于伦敦肯辛顿花园的蛇形画廊开始邀请世界著名建筑师每年一度在他们的产业上为画廊设计一个临时性展出装置。

2010 年的蛇形画廊展厅由法国建筑师让·努维尔设计，建筑的红色既呼应着伦敦的标志性颜色，又体现着建筑师对于周围绿调自然环境的现象学认知。

Since the year of 2000, the Serpentine Gallery, located in Kensington Gardens, London, has invited world-renowned architects to create temporary gallery pavilions annually on the site.

The gallery pavilion in 2010 was designed by Jean Nouvel; the red color of the pavilion echoed the signature color of the city, while also reflecting the phenomenological argument of the architect for natural green environment.

小型的"临时美术馆"是以较小的代价营造的实验性空间，它更新着人们对于当代艺术空间的看法，自身也是某种艺术品。

The small "temporary museum", with a comparatively low building cost, is an experimental space, which periodically refreshes people's vision of the art space, becoming an artwork in and of itself.

蛇形画廊展亭 2010 外景
An exterior view of the Serpentine Gallery Pavilion, 2010

2010 蛇形画廊展亭的可以折叠的红色屋顶
Foldable red roof of the Serpentine Gallery Pavilion, 2010

柏林国家博物馆（老博物馆和新国家画廊）
National Museum in Berlin
(Altes Museum and New National Gallery)

德国柏林
Berlin, Germany

老博物馆
Altes Museum

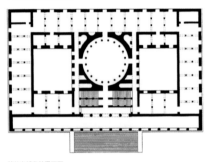

柏林老博物馆平面图
Plan of Altes Museum, Berlin

老博物馆位于柏林博物馆岛边缘，比邻柏林大教堂，博物馆的主要藏品为古希腊和古罗马文物，但建筑作品本身则透露出设计者卡尔·F·辛克尔重新诠释古典建筑的意图。在博物馆入口处，当人们面对巨大的阶梯拾级而上的时候，城市的水平街景为不断变化的柱间距所分割，产生的视差形成电影画格般的观看效果。这座新古典主义的博物馆由此成为新兴德国的国家象征，在其坚实的堡垒立面之上，也呈现出和城市交流的强烈动态。

Altes Museum is located on the Museum Island of Berlin and next to the Berliner Dom. The museum's main collections are cultural relics from ancient Greek and Roman times, but the building itself represents the will of the architect, Karl Friedrich Schinkel, to reinterpret classical architecture. At the foyer of the museum, when visitors climb up the giant steps to view the city, the horizontal streetscape, divided by the columns, forms a film-type spectacle due to parallax effect. Thus, the Neo-Classical museum has become a symbol of the rising German state, presenting a strong sense of motion interacting with the city, in addition to its solid, castle-like façades.

今天，在资本主义上升时期建立起的西方美术馆绝大多数已经老旧，因此它们的建筑也面临着自我更新的问题。

Today, most Western museums established in the rising period of Western capitalism have aged, and their architecture thus faces the issue of renovation.

柏林老博物馆立面
Façade of Altes Museum, Berlin

柏林新国家画廊
Neue National Gallery, Berlin

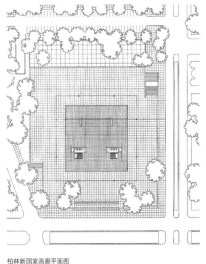

柏林新国家画廊平面图
Plan of Neue National Gallery, Berlin

0   20   60 m

20 世纪中叶落成的柏林新国家画廊位于德国柏林波茨坦广场，是现代建筑的先驱密斯·凡·德·罗向他的前辈辛克尔致敬的作品。这座 100 年后的建筑物虽然使用了完全不同的现代主义手法，但它的平面与立面、内部空间与城市的关系却与辛克尔的老馆如出一辙，都保持着承前启后的开放姿态。

Built in mid-20th century, and situated in the Potsdamer Platz, Neue National Gallery in Berlin is a tribute by Mies van der Rohe, a pioneer of modern architecture, to his predecessor Schinkel. Although 100 years later the building adopts a Modernist approach, different from Schinkel's, the relationship between the façade and the plan, as well as the interior and its urban environment, remains consistent with that seen in Schinkel's Altes Museum. Both have sustained traditions and opened a window to the future.

新型建造的方式"倒逼"美术馆功能使用的更新，是一种"自外而内"
的改变方式。

New types of construction can force art museums to refresh their
functions; this is a change going from the outside to the inside.

卢浮宫博物馆扩建
The Louvre Expansion

法国巴黎
Paris, France

卢浮宫博物馆首层平面图
Plan of ground floor, the Louvre

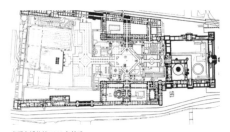

卢浮宫博物馆 1989 年扩建
The Louvre Expansion in 1989

卢浮宫改扩建工程，是 1989 年政府为纪念法国大革命 200 周年的巴黎十大工程之一。美籍华人建筑师贝聿铭设计的改扩建方案并没有显著地改动原建筑立面形象或平面结构，他著名的玻璃金字塔主要是创造了一个接纳更大人流的公共入口，并增加了大量可用的地下空间，缓解了展览空间的拥堵，从而间接地改变了博物馆的面貌。

The expansion of the Louvre is one of the ten "Grand Projects" in Paris initiated by the French government in 1989, in commemoration of the 200th anniversary of the French Revolution. Chinese-American architect I. M. Pei did not significantly change the façade or plan of the original museum; his renowned glass pyramid mainly created a public entrance to accommodate bigger flow of visitors, and added more usable underground space. This has successfully remedied the problem of crowdedness of display spaces and thus indirectly changed the museum's appearance.

以调整美术馆内部空间和品质来促进它潜在的变化，是"自内而外"的更新方式。

To promote potential changes by adjusting the interior space and improving its quality, this is the type of museum renewal "from the inside out".

卢浮宫博物馆玻璃金字塔入口内景
An interior view of the glass pyramid entrance to the Louvre

卢浮宫博物馆和玻璃金字塔外景
An exterior view of the Louvre and the glass pyramid

纽约新当代美术馆
The New Museum of Contemporary Art, New York

美国纽约
New York, USA

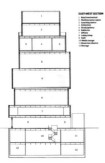

纽约新当代美术馆剖面图
Section of the New Museum of Contemporary Art, New York

0   5   15m

新当代美术馆位于纽约曼哈顿下城东区包厘街 (Bowery) 附近。该馆每年有六个主要的当代艺术展，以及五个主要媒体艺术展。建筑由六个没有分割的"盒子"堆叠而成，构成一体化的六个不同画廊空间。极简的建筑外观和未加预设的建筑平面呈现出新时代中艺术展览空间的多样性可能。

建筑的六个矩形盒子的每一个盒子都有不同的楼层面积和净高度，新颖结构设计确保展览空间从里到外没有任何的柱子。通过体量的错动解决采光和景观的问题，不同展厅之间呈现微妙的差别。

The New Museum of Contemporary Art is located in Manhattan's Lower East Side, near Bowery street. The museum holds six major contemporary art shows and five major media art exhibitions each year. The building consists of six undivided "boxes" stacked in a pile, forming six different gallery spaces with a uniform appearance. A minimalist façade and a building plan with no pre-setting represent diversified potentials of art display spaces in a new age. In particular, each of the six boxes has a different floor area and height. Innovative structural design ensures no columns throughout the display spaces. The stacking of the "boxes" offers solutions to lighting and landscape, resulting in subtle differences between the galleries.

新时代的美术馆如同它所承载的艺术活动一样，呈现出多样化和不拘一格的面貌，很难讲今天的美术馆设计还有什么"定则"。

Museums in the new age, like the art events they host, become diversified and unrestrained in appearance. Accordingly, it is hard to define rules for contemporary museum design.

新当代美术馆显著地改变了包厘街的街景。
The New Museum of Contemporary Art has a significant impact on the Bowery street scene.

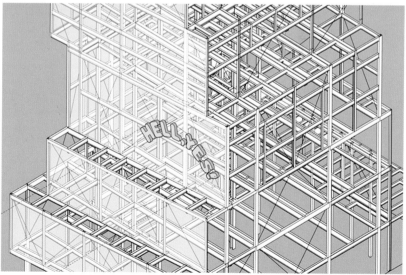

新当代美术馆建筑结构的分析
A structural analysis of the New Museum of Contemporary Art building

第三章

城市中的美术馆
Urban Museums

○ 美术馆塑造城市的外在形象，也承担城市的内部功能。
Museums form a city's external image and carry the city's internal functions as well.

○ 美术馆在城市中。
Museums are in the city.

○ 一座城市就是一座美术馆。
A city is a museum.

○ 反之，美术馆也是一座小型的城市。
Conversely, a museum is a city.

○ 美术馆的空间是城市肌理的映射和延续。
Spaces of a museum reflect and extend the fabric of a city.

○ 美术馆如今是一类重要的城市公共空间。
A museum today serves as an important type of urban public spaces.

○ 它兼有教育和娱乐的功能。
It is both pedagogical and recreational in functionality.

○ 美术馆常常成为城市设计的突破环节。
Museums often become a breakthrough in urban design process.

○ 建筑项目也应该合理考虑城市设计。
Architectural projects should reasonably take urban design into consideration.

美术馆的城市界面和形象
Urban Interfaces and Images of Museums

在密度较高的城市中，为了协调美术馆和城市整体形象的关系，除了控制建筑主体的高度外，对围绕主体建筑的局部构筑物和标识物也应做出规定；建筑物的形象应该注意在各个面都与周边已有建筑群关系协调，为此，各个立面不妨适当采用不同的设计手法。在有地形存在的城市区域，设计应该特别考虑各个建筑物地面层之间可能存在的高差，并注意如何使它们各自的立面共同形成各有特色又连续统一的景观界面。

In the densely populated urban areas, there should be specifications to control the height of the museum's main buildings, as well as annexations and signs surrounding it, in order to coordinate the relationship between the museum and the city. The museum building appearance in all its façades should be in harmony with existing building groups nearby; therefore, it is suggested that different façades adopt different designing approaches. For museums located in urban areas with prominent topography, possible height difference between different buildings of the museum needs to be considered in designing; attentions should be paid to unifying the façades with different features into a common interface that offers a continuous landscape.

立面体量控制
Controls of façades and masses

数字北京大厦
Digital Beijing Mansion

国家体育馆
National Stadium

北辰洲际酒店
Beichen
InterContinental Hotel

国家游泳中心
National Aquatics Center
("Water Cube")

MoMA 建筑体量控制
Building mass control of MoMA

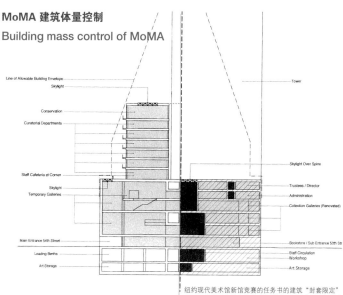

纽约现代美术馆新馆竞赛的任务书的建筑"封套限定"使得建筑设计和城市的关系中规中矩，同时又留有创造的余地。

The competition guideline for the new MoMA building sets an architectural "envelope", which mediates the relationship between the building and the city, while leaving room for creativity.

国家美术馆新馆（提议）的国际竞赛任务书所示的城市与建筑的地标关系
Relationship between the building and the city, as specified in Competition Guidelines for the proposed new National Art Museum of China

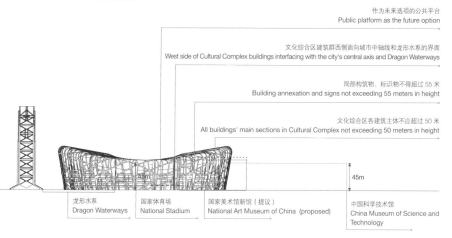

作为未来选项的公共平台
Public platform as the future option

文化综合区建筑群西侧面向城市中轴线和龙形水系的界面
West side of Cultural Complex buildings interfacing with the city's central axis and Dragon Waterways

局部构筑物、标识物不得超过 55 米
Building annexation and signs not exceeding 55 meters in height

文化综合区各建筑主体不应超过 50 米
All buildings' main sections in Cultural Complex not exceeding 50 meters in height

45m

龙形水系
Dragon Waterways

国家体育场
National Stadium

国家美术馆新馆（提议）
National Art Museum of China (proposed)

中国科学技术馆
China Museum of Science and Technology

城市就是美术馆
A City is a Museum

交通与美术馆

根据美术馆的不同城市语境和设计规模，城市交通与美术馆衔接的方式不尽相同。一部分美术馆设在城市主要通路的邻近，另一部分美术馆则有意识地远离交通要道。有的时候，城市交通可以成为美术馆特殊的"户外展览"设计的一部分。

城市环境造就了美术馆：其一，美术馆的用地性质往往是城市中比较特殊的所在；其二，如何引领周边城市区域的发展决定了一个美术馆项目的成败。

Transportation and the Museum

Based on urban context and museum scale, there are a variety of ways to connect a museum to the city. Some museums are in proximity to major urban thoroughfares while others are purposely distanced from traffic. Sometimes the urban transportation itself might become a special part of the museum, as an "outdoor display".

Urban environment makes a museum: first, the land use of a museum is often particular to its urban context; second, how to lead the development of the museum neighborhood determines the fate of a museum project.

美术馆的城市功能

大多数展厅定位为全开放的公共空间，需要容纳相对多的观众同时参观，而展厅又要求相对肃穆和安静，因此带来了某种意义上的功能矛盾。当代美术馆逐渐不回避这样的矛盾，除了保留展览空间的主要功能"看"展览，它也引入了城市生活特有的多样性和各种不同的"活动"："学习"强调观众积极参与美术馆的空间；"交流"给大众提供自我组织和彼此融合的机会；"娱乐"使人身心放松。最后，美术馆还提供喧闹的城市不常有的"沉思"的空白空间。

The Urban Functions of a Museum

Most galleries are established as fully open public spaces that can accommodate large audiences at once while maintaining solemnity and silence, thus engendering a functional paradox in certain sense. Contemporary museums gradually opt to face this paradoxical issue. In addition to preserving the major function of the gallery spaces for visitors to "view" exhibitions, they adopt the diversity specific to urban life and a variety of "activities" including: "learning", to emphasize the active involvement of the audience with the museum spaces, "communication", to provide opportunities for the general public to self-organize and socialize, and "recreation", to provide a leisure atmosphere. Last, museums also provide an unusual type of "void" space for contemplation, rarely found elsewhere in noisy cities.

美国纽约大都会艺术博物馆区位图
Location of the Metropolitan Museum of Art, New York, USA

N

0   30   90m

大都会艺术博物馆的北立面与中央公园自然环境的界面
Interface between the north façade of the Metropolitan
Museum of Art and the greenery of Central Park

具有公益性质的大都会艺术博物馆当初被破例允许在纽约中央公园内择址，它的东立面朝向公园大道，其他三个立面嵌入公园绿地之中。经过数次主要的改扩建，大都会艺术博物馆成为城市生活和自然环境的纽带。

As a nonprofit institution, the Metropolitan Museum of Art (the Met) was exceptionally allowed to situate in Central Park, New York. Its east façade faces Park Avenue; three other façades merge into the greenery of the park. After several major expansions and rebuilding, the Met has become a point of confluence of urban dynamics and natural environment.

美国纽约 现代美术馆 区位图
Location of the Museum of Modern Art (MoMA), New York, USA

N
0   10   30m

从东向西看现代美术馆和整个街区
East-west view of MoMA and the entire block

纽约现代美术馆处在紧致的城市街区中。因此，从视觉上和心理上贯通 54 街和 53 街就成为几代设计师的共同任务。虽然私人美术馆不可能做到完全开放，但纽约还是作为一种语境被引入现代美术馆的室内空间中。

MoMA finds itself in a compact urban block. As a result, channeling 54th and 53rd Street has become a shared challenge for generations of architects. Although this private museum is unable to be completely open to the city, New York nonetheless "inhabits" the inner spaces of MoMA as its context.

法国巴黎 奥赛美术馆 区位图
Location of Musée d'Orsay, Paris, France

N
0   100   300m

法国巴黎 蓬皮杜艺术中心 区位图
Location of Le Centre Pompidou, Paris, France

N
0   100   300m

美国堪萨斯城 纳尔逊－阿特金斯艺术博物馆及其新馆 区位图
Location of the Nelson-Atkins Museum of Art and its expansion, Kansas City, USA

N
0   30   90m

美国沃思堡 金贝尔美术馆 区位图
Location of Kimbell Art Museum, Fort Worth, USA

N
0   30   90m

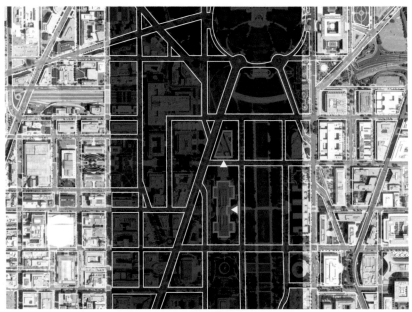

美国华盛顿特区 国家艺术馆及其东馆 区位图
Location the National Gallery of Art and its east wing, Washington D.C., USA

N 0 30 90m

美国辛辛那提 罗森塔当代艺术中心 区位图
Location of Rosenthal Center for Contemporary Art, Cincinnati, USA

N 0 30 90m

西班牙墨里达 国家罗马艺术博物馆 区位图
Location of National Museum of Roman Art, Mérida, Spain

N 0 30 90m

西班牙毕尔巴鄂 古根海姆美术馆 区位图
Location of Guggenheim Museum, Bilbao, Spain

N 0 30 90m

英国伦敦 大英博物馆 区位图
Location of the British Museum, London, UK

N

0   100   300m

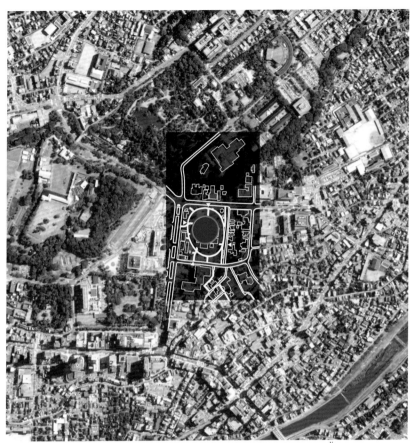

日本金泽 21 世纪当代美术馆 区位图
Location of the 21st Century Museum of Contemporary Art, Kanazawa, Japan

N   0   100   300m

公共聚会
Public Gathering

当代美术馆不仅是"人看画"的地方，而且也日益变成"人看人"的场所。无论室内还是室外，应鼓励营造气氛适宜、便于管理的公共聚会场所。但要注意不对美术馆室内使用和安保措施造成干扰。结合美术馆和城市的特点，适合公共聚会的空间既可以是中等规模的城市广场，也可以是私人尺度的小型封闭空间。为了公共聚会的灵活性起见，美术馆户外不宜设置纪念性过强的大型广场，美术馆附属的户外公共空间应当既是美术馆室内公共空间的延伸，也是周边城市地区现有公共空间的补充。

Contemporary museums are not merely about "viewing art" but also about "viewing people". It is encouraged to operate public venues, inside or as annex to a museum. These public spaces are supposed to be in line with the atmosphere of the museum and easy to manage, and not to disrupt the interior uses of the museum or security control. Spaces suitable for public gathering could be mid-scale urban plazas or human-scale small enclosures, depending on the museum and the city conditions, For the sake of flexibility in public gatherings, it is not suitable for a museum to set a large-scale outdoor plaza dominated with monumental features. Public spaces as annexes to the museum should serve as an extension to its interior public spaces, as well as a complement to existing public spaces of the surrounding urban areas.

1　大都会艺术博物馆公共等候区
　　Public waiting area at the Metropolitan Museum of Art

2　纽约现代美术馆的公共聚会空间
　　Public meeting space in MoMA, New York

公共娱乐
Public Recreation

传统的美术馆通常排斥展出空间的娱乐功能。当代的美术馆则将娱乐休闲看成是美术馆城市功能的一部分，这些功能受到专业人员之外的更广大城市公众的欢迎。美术馆附属的餐饮、观演等设施在完成自身任务之余，在不干扰美术馆自身使用的前提下，可以成为城市公共娱乐设施的一部分。美术馆的门禁、管理等系统应做相应的设计和调整，以创造出一部分和城市无缝对接的娱乐功能。

Traditional museums often reject the recreational function of display spaces. In contrast, contemporary museums take recreation as part of its urban function, which is desired by a larger urban population beyond professionals. After finishing serving the museum's visitors, the museum's restaurants, theaters, and other facilities can consider serving the larger urban population, becoming part of the public recreational facilities of the city on the condition of not interfering with the museum's operation. In order to do so, systems of museum management and security should be designed or adjusted accordingly to create a recreational function smoothly accessible to the city's general public.

1　上海世界博览会德国馆内的即兴演出
　　Improvisation at German Pavilion,
　　2010 Shanghai Expo

2　费城美术馆公共空间中的音乐演出
　　Concert in the public space of
　　Philadelphia Museum of Art

117

公共教育
Public Education

美术馆的公共教育既是自上而下的"教谕",也是自下而上的"学习",后者需要更方便地进入和使用,美术馆提供的教室、图书馆、剧场等设施应该有意识地向观众以外的一般大众开放。和公共娱乐设施一样,美术馆的门禁、管理等系统应该为这部分公共空间做相应的设计调整,以便和城市无缝对接。尤其值得注意的是,美术馆的户外展示空间也应当提供积极的公共教育功能。

Public education in museums refer to both a top-down approach to "education" as well as a bottom-up approach to "learning", while the latter requires easy access and easy use. Classrooms, libraries, and theaters of the museum should consciously provide access to the general public in addition to the standard museum audience. The museum security and management should be designed or adjusted accordingly in public spaces for the museum education function, as in the case of public recreational facilities. In particular, outdoor display spaces of the museum should actively serve the function of public education.

1   大都会博物馆的公共教育活动
     Public education events at the Metropolitan
     Museum of Art

2   同学们围拢在中国美术馆志愿者周围,倾听她讲述
     "那过去的故事"
     A NAMOC volunteer tells the "story of the
     past" to the students around her.

3   艺术之旅:走进中国美术馆长廊
     A trip to art: strolling down the NAMOC veranda

城市设计中的美术馆
Museum in Urban Design

美术馆是城市建设的宠儿

现代的规划体制常常把美术馆建筑看成"填补"城市肌理的一部分，它虽然吸引眼球却和周边的城市环境没有太多的互动和往来，其实美术馆也可以能动地影响城市设计。根据美术馆和周边环境的不同关系，美术馆建筑可以带动街区的发展和开放空间互补，与城市生活立体交叉，或是在后续开发中逐渐"融入"乃至"吸纳"邻近的城市肌理。美术馆周边的城市环境也是美术馆。

Museum is the Favorite to Urban Development

Modern planning often relegates the museum buildings as complementary part of the urban fabric, with eye-catching features but little interaction with their urban surroundings. Actually, museums can actively influence urban design. Depending on the type of relationships between the museum and its urban context, the museum has the potential to lead the development of the urban block, to complement its public space, to interact with urban life, or to "fuse" with and even "absorb" nearby urban fabric during follow-up development. The urban environment surrounding the museum is also part of the museum.

- 拱卫式

  以美术馆为中心，带动附属的城市公共空间由内向外地发展。

  Derivative

  Museums, as the core, lead the development of affiliated urban public spaces from the inside out.

- 顺延式

  美术馆和它所影响的城市公共空间是对应、反映关系。

  Extension

  Urban spaces mirror the museum space as their counterpart or reflection.

- 交叉式

  通过建筑类型的交叉与叠加，城市公共空间和美术馆可以立体混合。

  Crossover

  Particular building typology makes the public space overlapped or interwoven with museum space, mixing their uses in a three-dimensional fashion.

- 演进式

  美术馆成片占有城市土地之后，再由外而内地发展出内部的公共空间。

  Evolution

  Museums claim the urban land in bulk and nourishes its internal public space from the outside in.

公共开放空间

采用"整存整取"的方法，蓬皮杜国家艺术文化中心成为巴黎中心区新古典肌理中的一处例外。由建筑集成一处的文化功能和留白而成的广场开放空间互相补充，有效地吸收本区域的人流，成为有活力的公共生活的引擎。

Public Open Space

In a strategy of "whole in, whole out", Le Centre Pompidou became an exception to the Neo-Classical urban fabric of central Paris. The cultural functions integrated by the building and the plaza space deliberated left open complement each other, together effectively absorbing pedestrian flow of this area and becoming the engine of energetic public life.

拱卫式
Derivative

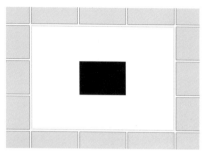

■ 美术馆建筑
Museum space

□ 城市公共空间
Public spaces

▨ 附近城市肌理
Nearby urban fabric

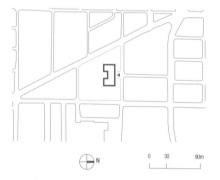

⊕ N   0   30   90m

金贝尔美术馆（美国德克萨斯）
Kimbell Art Museum, Fort Worth, USA

顺延式
Extension

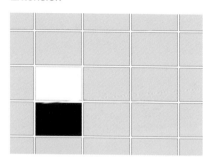

■ 美术馆建筑
Museum space

□ 城市公共空间
Public spaces

▨ 附近城市肌理
Nearby urban fabric

⊕ N   0   100   300m

蓬皮杜中心（法国巴黎）
Le Centre Pompidou, Paris, France

交叉式
Crossover

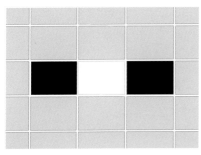

演进式
Evolution

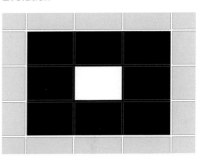

■ 美术馆建筑
Museum space

□ 城市公共空间
Public spaces

▨ 附近城市肌理
Nearby urban fabric

■ 美术馆建筑
Museum space

□ 城市公共空间
Public spaces

▨ 附近城市肌理
Nearby urban fabric

N

| 0 | 10 | 30m |

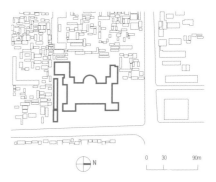

N

| 0 | 30 | 90m |

现代美术馆（美国纽约）
MoMA, New York, USA

中国美术馆（中国北京）
NAMOC, Beijing, China

改变城市的美术馆
Museums which Change the City

通常的美术馆建筑设计不考虑，或不重视建筑师在建筑单体之外的设计工作中所起的作用。事实上，除了完成建筑物自身的设计工作外，对整个区域建筑物用地红线之外的公共活动空间的使用情况，建筑师最好能也做出合理的预期。配合建筑设计最好有相应的城市设计导则，着重强调整体规划与单体设计之间的互动和促进。

建筑师应在单体方案设计的同时考虑公共活动区域的设计，并与业主及时交流沟通，以确保公共活动区域的设计与单体建筑的设计能够有效衔接。在新开发地段，建筑设计应预先考虑公共空间的位置、形态、景观效果以及对建筑物立面效果、尺度关系的影响，形成初步概念，以利后续景观设计与扩初设计同步进行。

Most museum designs rarely consider or even disregard the role architects play in design work beyond the museum. In fact, in addition to attending to the museum building design itself, architects should reasonably take into consideration the use of public spaces in the whole neighborhood beyond the red line of construction. Ideally, urban design guidelines should be compiled to guide museum design, with an emphasis on mutual promotion and interaction between museum building and area planning.

When designing for museum buildings, architects should consider annexed outdoor spaces for the public. Through prompt communication with clients, architects can ensure the effective connection between their immediate projects and the long-term public uses of the space. In new developments, architectural designs should take into consideration the location, the form, and the landscape of public spaces, as well as the impact of public spaces on the architectural façades and the scale of the museum. This preliminary conception process will benefit the following landscaping projects and detailing designs.

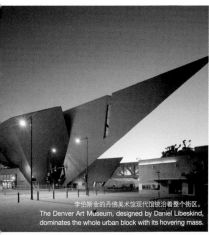

李伯斯金的丹佛美术馆现代馆统治着整个街区。
The Denver Art Museum, designed by Daniel Libeskind, dominates the whole urban block with its hovering mass.

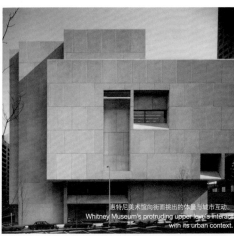

惠特尼美术馆向街面挑出的体量与城市互动。
Whitney Museum's protruding upper levels interact with its urban context.

奥克兰战争纪念博物馆坐落于城市郊区的自然环境中。
The Auckland War Memorial Museum finds itself in a natural environment within a suburban area.

卢浮宫的金字塔室内约等于城市公共空间，比如商场空间。
The underground area of the Louvre Pyramid is equivalent to urban public space such as shopping malls.

大都会艺术博物馆门前的大台阶
Big steps at the entrance of the Metropolitan Museum of Art

美国亚利桑那州立大学的画廊向大街开放。
The art gallery of Arizona State University interfaces with the street.

第四章

建筑形象的问题
Image of Museum Building

○ 建筑形象是专业人员和公众分歧众多的方面。
Professional expertise often diverges a lot from public opinions on museum image.

○ "形象"是美术馆最重要的设计问题吗？
Is the "image" of a museum the most important issue in design?

○ 讨论设计问题，意味着寻找共识而非突出差异。
Discussion on design is intended to seek consensus instead of disagreement.

○ 全方位的设计、切题的设计和最大效益的设计
Holistic design, context-driven design and optimized design

○ 贴近城市的形态
Architectural forms that are set closely in urban context

○ 因地制宜的色彩
Colors that are locally-adaptable

○ 便于交流的空间
Spaces that are convenient for communication

○ 合乎使用的尺度
Scales that are suitable for human uses

○ "中国"的空间 vs 贴近中国艺术的空间
"Chinese" space vs. space for Chinese art

美术馆的形象
Image of Museum Buildings

- 对于美术馆的主管部门、美术馆的设计师、美术馆的实际使用者以及一般公众而言，建筑形象往往意味着截然不同的东西；
- 对美术馆的设计与使用而言，"形象"只是诸多关注中的一个方面；
- 当讨论美术馆的"形象"问题时，可以讨论的部分是那些能够寻找到共识的问题。

- To the museum management, designers, actual users, and the general public, the image of the museum often indicates drastically different meanings;
- Among the many aspects to consider for museum design and function, "image" is just one.
- When discussing on the museum "image" issues, we need to focus on topics which can lead to consensus instead of disagreement.

建筑物是美术馆最大的艺术品。

伫立在原慕尼黑城城门之前的慕尼黑雕塑美术馆，是巴伐利亚的国王路德维希为了他心目中"德国的雅典"而建造的。建筑师利奥·冯·克伦泽构想过罗马式、文艺复兴式的不同方案，最终确定的建筑方案的三个立面含有十八个壁龛，里面陈列着希腊罗马的雕塑原作，如此建筑的外表自身便是一座蔚为大观的城市露天"画廊"。

Museum Building Itself is the Largest Exhibit.

The Glyptothek standing in front of the original city gate of Munich, Germany, was built by Ludwig Otto Friedrich Wilhelm of Bavaria as "German Athens" in his mind. The architect Leo von Klenze had envisioned Roman and Renaissance styles for the museum. In the finally chosen design, eighteen niches on three façades displayed the original works of Greek and Roman sculpture. A museum building in such appearance itself is a grand outdoor "gallery" for the city.

设计的约定性原则
The Customary Principle of Museum Design

004 城市 Urbanism 312 一点哲学 My Philosophy
320 国家美术馆新馆（提议）建筑竞赛第一阶段任务书摘要
Guideline Excerpts from the First Round Competition of the Proposed NAMOC New Building

很难为千变万化的建筑设计寻求截然不变的原则，在此我们讨论的是既有约束性也有约定性的原则：

- 应该整体和系统地考虑一座美术馆的设计；
- 没有最好的设计，只有最合乎情境的、最对题的设计；
- 设计要以最少的人力、物力消耗，去获取最大的收益，避免不必要的耗费。

It's almost impossible to chase after a definite design rule for drastically different buildings. The rules we discuss are restrictive and customary as well.

- We should consider the museum design as a whole and systematically.
- There are no best designs, but most suitable and most relevant ones.
- We must seek to maximize gains while minimizing the consumption of manpower and material resources, to avoid unnecessary waste.

1. 我看起来就是一幅画，好在就不会有滑稽的"联想"了。
   I look like a painting, fortunately with no other funny "association".

2. 我是一套建筑语言，很严谨，很无趣。
   I am a set of architectural grammars, strict but uninteresting.

3. 我很抽象，但我可以促进观众和建筑之间"看来看去"的交流。
   I am abstract, but I am a window that could promote the "back and forth" communication between the audience and the building.

4. 我很生动，但我看起来不像是幢建筑……
   I am vivid, but I look almost not a building...

5. 我是一个大写的建筑符号。
   I am a capitalized architectural sign.

"形象"的不同含义

大多数建筑师，甚至那些建造了非常成功的建筑作品的建筑大师，其实不怎么关心建筑在普通人心目中的形象——建筑的形象"看上去"往往有不同的含义。在这种情况下设计者和使用者必须就"形象"的定义取得起码的共识。

Various Meanings of the Museum Image

Most architects, even those who are known for their master pieces, rarely care about ordinary people's opinions on architectural images. In fact, the meanings of the image of a building appear to be different. In such a case, the designer and the user must arrive at a minimum consensus concerning the ideal image of a museum.

形态
Form

建筑平面尽量遵从现有规划的地块形状，充分融入城市语境，提高基地面积利用效率。

● 考虑到公共建筑通常巨大的尺寸，建筑内部平面也应形成清晰的结构关系，特别是深入理解建筑功能的特定关系，合理地布置美术馆主要的空间区块，构成自然的外部形态；

● 应该现实地看待建筑和城市的关系，一座美术馆不可能不切实际地要求其他建筑物服从自己的需求，相反，应该尽可能地利用自身灵活的形态特点，在建筑内部或邻近区域，创造出合乎要求、面向城市的公共空间。

The building plan should respect the existing land planning and merge fully into the urban context, to increase the efficiency of its land use.

● In consideration of the usually gigantic scale of public buildings, the interior plan needs to be structured clearly, with a deep understanding of the particular relevance to various interrelated building functions. The reasonable distribution of the major spatial elements naturally forms the exterior appearance of the building.

● We should approach the relationship between the museum and the city realistically. A museum cannot require unreasonably neighboring buildings to conform to its own needs; instead, it should flexibly adapt its form to create public spaces within and near its building that meet requirements and are open to the city.

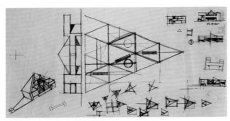

贝聿铭对三角形建筑平面的绘图推敲
Sketches by I. M. Pei on the would-be triangle museum plan

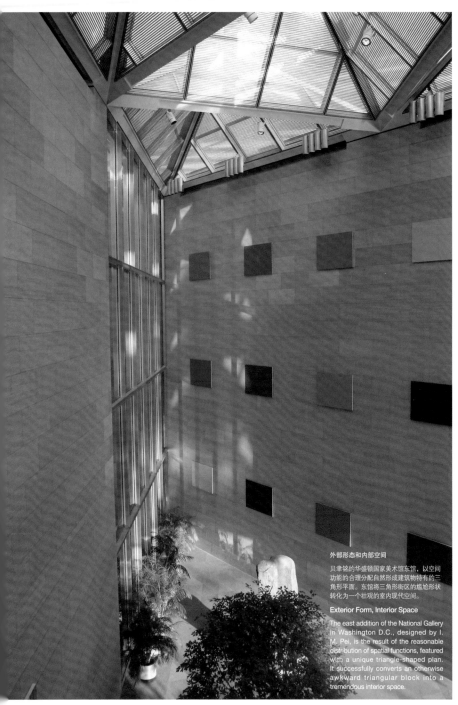

外部形态和内部空间

贝聿铭的华盛顿国家美术馆东馆，以空间功能的合理分配自然形成建筑物特有的三角形平面。东馆将三角形街区的尴尬形状转化为一个壮观的室内现代空间。

Exterior Form, Interior Space

The east addition of the National Gallery in Washington D.C., designed by I. M. Pei, is the result of the reasonable distribution of spatial functions, featured with a unique triangle-shaped plan. It successfully converts an otherwise awkward triangular block into a tremendous interior space.

方形

方形或者矩形平面的美术馆对应着规则的建筑平面，很容易容纳传统类型的画廊空间，构成直来直去的展览流线，基本不造成空间的浪费和死角。

圆形

圆形或曲线平面美术馆有利于造就一种环绕观众的全方位展出环境，形成整体有聚合力和亲和力的公共空间，但分隔墙较难设置，不利于需要直墙的平面类型美术品（例如绘画）和那些需要多个单独展出环境的美术品（例如暗室投影）。

Square

Museums with rectangular or square plans correspond to ordered building surfaces, easily accommodating traditional types of gallery spaces and forming straight visitor routes with little wasted space and few dead corners.

Circular

Museums with circular or curvy plans facilitate a panoramic environment to embrace the audience, resulting in an integrated public space with strong affinity. However, it is a bit awkward to set spatial divisions in such spaces. These spaces are also difficult for 2-D works demanding flat surface (e.g. paintings), as well as for objects and installations requiring separate enclosures (e.g. projection in dark room).

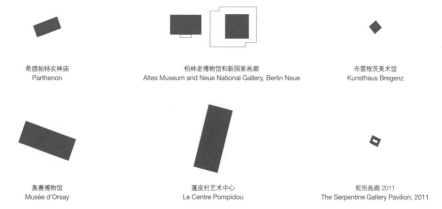

希腊帕特农神庙
Parthenon

柏林老博物馆和新国家画廊
Altes Museum and Neue National Gallery, Berlin Neue

布雷根茨美术馆
Kunsthaus Bregenz

奥赛博物馆
Musée d'Orsay

蓬皮杜艺术中心
Le Centre Pompidou

蛇形画廊 2011
The Serpentine Gallery Pavilion, 2011

异形

异形平面的美术馆在建筑外部和内部都造成有强烈动态的空间，形成展品和环境有机结合的气氛，适于富于创意的美术馆设计和多变起伏的地形。

Irregular

The museums with irregular-shape plans generate dynamic outdoor and indoor spaces, and combine art objects with the environment in an organic way. This type is suitable for uneven topography or creative projects.

伦敦水晶宫
The Crystal Palace, London

金泽 21 世纪当代美术馆
The 21st Century Museum of
Contemporary Art, Kanazawa

毕尔巴鄂古根海姆美术馆
Guggenheim Museum,
Bilbao

蛇形画廊 2012
The Serpentine Gallery
Pavilion, 2012

蛇形画廊 2009
The Serpentine Gallery
Pavilion, 2009

组合形

以上的平面类型在同一建筑物内部的聚合和叠加。

Combined

The juxtaposition and overlay of aforementioned building plans

大英博物馆
The British Museum

乌菲齐美术馆
The Uffizi Gallery

大都会艺术博物馆
The Metropolitan Museum of Art

芝加哥艺术学院博物馆
The Art Institute of Chicago

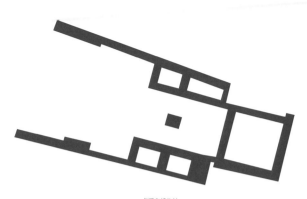

卢浮宫博物馆
The Louvre

纽约现代美术馆
MoMA, New York

泰特现代艺术馆
Tate Modern

美秀艺术馆
Miho Museum

万神殿
Pantheon

国家罗马艺术博物馆
National Museum of Roman Art

纽约新当代美术馆
The New Museum of
Contemporary Art, New York

金贝尔美术馆
Kimbell Art Museum

辛辛那提当代艺术中心
Rosenthal Center for Contemporary Art

蛇形画廊 2010
The Serpentine Gallery Pavilion, 2010

分散形

美术馆除了可以是自成一体的"独栋"建筑物，还可以是不同类型平面的松散组合。常见于新老美术馆群组，或新兴美术馆群落。

Complex

Instead of a self-contained structure, a museum can be a complex with a group of units organized loosely. This type of plan is often seen in the old-and-new museum pairs, or avant-garde museum campus.

纳尔逊 – 阿特金斯艺术博物馆
The Nelson-Atkins Museum of Art

华盛顿国家艺术馆
National Art Museum in
Washington D.C.

北京尤仑斯当代艺术中心
UCCA Center for Contemporary Art

盖蒂中心
The Getty Center

色彩
Color

从不同美术馆方案的色彩比较和材质研究使我们认识到:

● 复杂的地理因素决定了特定色调材质的建筑适用与否。例如，中国北方地区的尘霾天气将影响到大片使用浅色易脏材质的建筑表现;

● 由于公共建筑过大的尺度和不同构件间的差异，个体色彩的总和不等于它们间的简单相加，因此，只从模型出发的建筑色彩设定容易显得简化、单调，在现实中难以兑现;

● 在具体环境、时刻中，结合不同的反光度和材质，在不同时间、天气和季节的光线下，会有不同的色彩差异。一部分原本过于繁复的色彩设定在建成后显得更加怪诞、紊乱，原本规整的大片幕墙玻璃在特定光线和装置条件下也会出现"去平滑"而产生视觉差异。

The comparison of various museum color schemes and the research on their textures give us the following findings:

● Complicated geographic factors affect the performance of color palette and thus determine their usability. For instance, the dusty climate of northern China has an enormous impact on buildings with light color schemes, making them vulnerable to pollution.

● The interplay between different components of an oversized public building contributes to a coherent overall color effect, larger than that from individual colors added up. Therefore, color schemes based only on computer models often appear simplistic, monotonous, and inapplicable in reality.

● The same color or texture will inevitably experience variety of appearances depending on lighting and reflection of the environment, the time, the weather, and the season. Overcomplicated color schemes will appear bizarre and disorderly when applied in reality. Even uniform glass façades, under different lighting conditions and with particular installation, may appear rippled or uneven visually.

	色彩 Colors	材质 Materials
帕特农神庙 Parthenon		
万神庙 Pantheon		
乌菲齐美术馆 The Uffizi Gallery		
大英博物馆 The British Museum		
大都会艺术博物馆 The Metropolitan Museum of Art		
柏林国家博物馆 National Museum, Berlin		
奥赛博物馆 Musée d'Orsay		
纳尔逊 – 阿特金斯艺术博物馆 The Nelson-Atkins Museum of Art		
美国国家艺术馆 The National Gallery of Art, Washington D. C.		
卢浮宫博物馆 The Louvre		
国家罗马艺术博物馆 National Museum of Roman Art		
芝加哥艺术学院博物馆 The Art Institute of Chicago		
纽约现代美术馆 MoMA, New York		
纽约古根海姆美术馆 Guggenheim Museum, New York		
金贝尔美术馆 Kimbell Art Museum		
蓬皮杜艺术中心 Le Centre Pompidou		
鹿特丹康泰美术馆 The Kunsthal, Rotterdam		
盖蒂中心 The Getty Center		
布雷根茨美术馆 Kunsthaus Bregenz		
毕尔巴鄂古根海姆美术馆 Guggenheim Museum, Bilbao		
美秀美术馆 Miho Museum		
泰特新馆 Tate Modern		
辛辛那提当代艺术中心 Rosenthal Center for Contemporary Art		
金泽 21 世纪当代美术馆 The 21st Century Museum of Contemporary Art, Kanazawa		
纽约新当代美术馆 The New Museum of Contemporary Art, New York		
伦敦水晶宫 The Crystal Palace London		
北京尤伦斯当代艺术中心 UCCA Center for Contemporary Art, Beijing		
蛇形画廊 2009 The Serpentine Gallery Pavilion, 2009		
蛇形画廊 2010 The Serpentine Gallery Pavilion, 2010		
蛇形画廊 2011 The Serpentine Gallery Pavilion, 2011		
蛇形画廊 2012 The Serpentine Gallery Pavilion, 2012		

材质
Materials

非反光材质
Non-reflective

石膏墙、混凝土、粗面石材、木材等构成的环境相对低调，容易和艺术品区分开来，广泛地使用于美术馆的室内装修，在建筑外部则形成与美术馆较为相称的素朴气质。

An environment composed of plaster, concrete, rustic rock and wood appears understated and easily discernible from art objects. Such non-reflective materials are used widely in the interior of museums. If applied to the exterior of buildings, they produce a simple feel, matching the character of art museums.

1 **混凝土**

在东方建筑的装修传统中，裸露的清水混凝土表面容易被视作"未完成"，但低调的混凝土材质可以形成均匀的视觉背景，比较适合现代风格的艺术陈设，用在建筑外部则形成与美术馆相符的优雅气息。

Concrete

In the East Asian tradition, unadorned concrete surfaces are often viewed as "incomplete," but low-key concrete textures can form an even visual backdrop appropriate for contemporary art displays. Today it is taken as an elegant building material for museum exteriors.

2 **石材**

理查德·迈耶为他设计的罗马的和平祭坛博物馆选择了淡褐色、粗面的石灰华作为外观建材，刻意以粗糙的质感衬托建筑久负盛名的古典环境，但是受到尺寸的限制石材会带来更多的分缝，用在美术馆内部，对艺术品产生一定的视觉干扰。

Stone

Richard Meier chose rustic travertine of light brown color for his Ara Pacis Museum in Rome, to provide a backdrop for its prestigious urban context that is unmistakably classical. But the stone claddings have size limitations. When applied inside the museum, stones with grids of lines will be inevitably prominent in sight, which might visually distract art displays.

3 **木质贴面**

长期以来一直是欧洲室内装饰的常见作法，也大量地出现在传统风格的画廊里，但是这种木质材料常与某些艺术风格关联，因此在现代艺术为主的展出环境并不那么常见。

Wooden Veneer

For a long time, the wooden veneer was commonly seen in European interior designs as well as in traditional galleries, but this wooden material is highly suggestive of particular art-forms and thus has fallen into disfavor with contemporary art galleries.

透明材质
Transparent

玻璃加上钢铁框架创造出采光优异的美术馆空间，带来了美术馆建筑设计的革命。透明材质意味着建筑新的外部形象同时也反映它不同的内部使用，与此同时，难以预料的眩光和反射光也对密闭环境中的艺术品展示条件提出了巨大挑战。

Glass in combination of its supporting steel frames facilitates well-lit museum spaces, bringing about a revolution for museum designs. Transparency means a new perception for architecture appearance which also reflects its different internal uses. Meanwhile, unexpected glare and reflection challenge the classical model of art display that favors a completely closed museum environment.

玻璃

玻璃满足了现代人视觉和心理上对于"透明性"的需求，有效地保护了公共环境下的艺术品。另一方面，这种"透明性"是不彻底的，玻璃界面不同角度的观感不同，并带来安全隐患，引进自然光则需要根据美术馆通常所要求的照度、湿度增加额外措施。

Glass

Glass meets the needs of modern people for visual and psychological transparency, protecting art works with effective insulation. On the other hand, the transparency of glass is often compromised, with views varying with different angles. Unheeded glass may bring potential safety issues for careless audience. If exposed in daylight, the glass cases will demand extra measures for luminance and humidity in keeping with museum standards.

反光材质
Reflective

反光材质的反射表面，复制和改变了美术馆的室内室外环境。在室外它创造出生动而多变的美术馆形象，在室内则带来难以忽视的光学效果，对于传统的展览空间而言并不常见。

Reflective surface mirrors and changes the visual appearance of museum space, creating a vivid and varying ambivalent museum image in its exterior as well as difficult-to-ignore optical consequences in its interior that may disrupt the normal art displays. Reflective materials are not commonly seen in traditional exhibition spaces.

钛合金

毕尔巴鄂古根海姆美术馆由曲面块体组合而成，内部采用钢结构，外表覆以闪闪发光的钛金属镀层。美术馆建在水边，钛金属镀层反射着水光天色，嵌入城市肌理的建筑平面使造型独特的博物馆成为城市的一部分，成为城市天际线上难以忽略的视觉焦点。

Titanium Alloy

The titanium-gilded Guggenheim Bilbao, standing by a river, is composed of curvy surfaces and steel internal structure that deliver the reflection of the environment of the sky and the river. The building plan woven in the urban fabric makes the unique-shaped museum part of the cityscape and an obvious point of focus in the Basque vista.

便于交流的空间
Space for Convenient Communication

很多时候"形象"只是美术馆空间设计的结果而非其原因

怎样为多样性的美术馆形象设计找到一种建设性的思路？抛开那些主观性较强的分析方法，我们粗略地把美术馆空间划分为公共空间、展览空间和办公空间三个层次，不同的设计决定了这三个空间层次叠加和组合的方式。优秀的设计让三个空间层次有机地融合、方便地交流，有利于美术馆最终的使用。

On many occasions, the museum image is merely the result of architectural space designs, instead of the goal.

How could we find a constructive means of assessing the quality of this image in a variety of contexts? Besides purely subjective analysis, we can roughly divide the museum into three categories including public space, exhibition space, and office space. Different designs combine and overlay these spaces in various ways. Good designs often tie these together organically for convenient coordination to serve the museum's needs.

叠加
纽约新当代美术馆空间结构

Overlay
Spatial structure of the New Museum
of Contemporary Art, New York

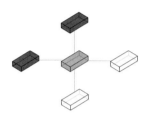

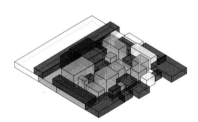

混合
金泽 21 世纪当代美术馆空间结构

Mixture
Spatial structure of the 21st Century Museum of
Contemporary Art, Kanazawa

公共
Public Space

展览
Exhibition Space

办公
Office Space

建筑各部分具有相同或相关属性的空间
金贝尔美术馆 / 沃斯堡 / 路易·康

Similar or correlated spaces forming
one building
Kimbell Art Museum, Fort Worth, Louis
Kahn

一个强有力的、统一个别、并使之成为整体的结构
布雷根茨美术馆 / 奥地利 / 卒姆托

One powerful and encompassing structure
that unifies all individual elements
Kunsthaus Bregenz, Austria, Peter Zumthor

一个防止过于生硬的区分和区域 的结构
劳力士学习中心 / 洛桑 / 妹岛和世、西泽立卫

A structure that avoids angular or rigid
dividing lines and division
Rolex Learning Center, Lausanne,
SANNA

有边界的庭院
金泽 21 世纪美术馆 / 金泽 / 妹岛和世、西泽立卫

Bordered Courtyards
The 21st Century Museum of
Contemporary Art, Kanazawa, SANNA

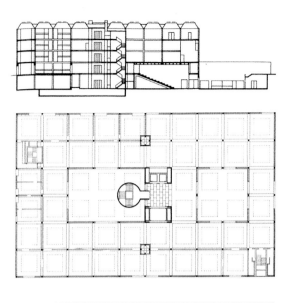

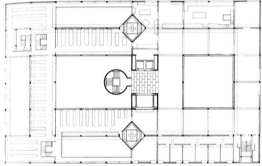

紧凑，避免死角和浪费的造型
耶鲁英国艺术中心 / 纽黑文 / 路易·康

Compact forms with less dead corners and wasted spaces
Yale Center of British Art, New Haven, Louis Kahn

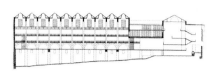

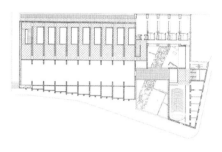

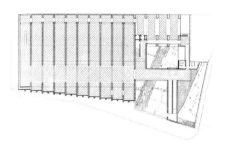

留白
波恩美术馆 / 波恩 / 亚克塞尔・舒特斯

Reserved Void
Bonn Museum of Art, Bonn, Axel Schultes

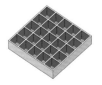

重复同构
国家罗马艺术博物馆 / 墨里达 / 拉斐尔・莫内欧

Repetition of Units
National Museum of Roman Art, Mérida, Rafael Moneo

路径即建筑
古根海姆美术馆 / 纽约 / 弗兰克·劳埃德·赖特

Path That Makes Space
Guggenheim Museum, New York, Frank
Lloyd Wright

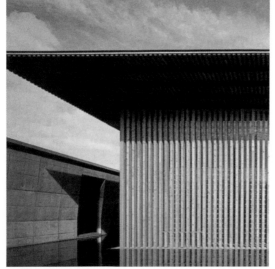

亭阁中心和周边对比
光明寺 / 西条市 / 安藤忠雄

Pavilion Center and its Periphery
Komyo-ji Buddhist Temple, Saijyo, Tadao
Ando

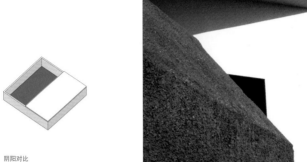

阴阳对比
舒拉格美术馆 / 巴塞尔 / 赫尔佐格和德默隆
Yin and Yang
Schaulager, Basel, Herzog & de Meuron

以上这些原则并非机械分割单独使用，相反，只有综合考虑本书中的各项内容才能产生优美的空间形态，富于感染力的空间形象，会"说话"的空间语言。归根结底，美术馆的建筑形象要求它的环境和功能一并讨论。

The aforementioned principles cannot be applied in a merely mechanical, rigid manner. Instead, a comprehensive approach that integrates the various perspectives of this book would help to produce beautiful spatial forms, impressive architectural images, and a building that can "speak" in a spatial language. As the bottom line, the architectural image of a museum can only be discussed in light of its environment and functions.

美术馆的尺度亦大亦小
Big Gallery, Small Gallery

当代艺术有尺度爆炸的趋向，但是美术馆需要兼顾极大和极小两种尺度。视觉印象和公共使用不妨极大化，人际的使用应有舒适的尺度。如何创造性地解决大小空间过渡的问题？中国传统城市给出了大小空间尺度营造和转换的若干实例，可以避免美术馆一味"巨构"的弊病。

- 住宅：小尺度单元
  住宅使用的尺度总是宜人舒适的，但是不适合大规模的公共使用。
- 公共空间：小尺度单元的放大
  在城市环境中，公共空间的大尺度不仅仅取决于高度，也取决于它在水平方向的延展；
- 园林：大尺度内部的小尺度间隔
  通过将室外和室内有机地渗透在一起，柔和地处理两者的边界，园林成功地解决了大小两种不同尺度之间的矛盾。

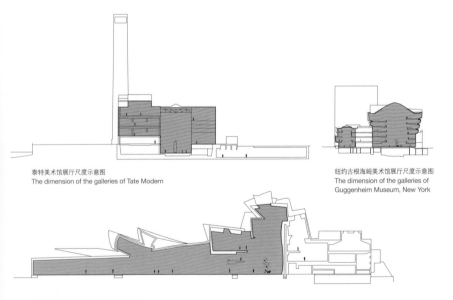

泰特美术馆展厅尺度示意图
The dimension of the galleries of Tate Modern

纽约古根海姆美术馆展厅尺度示意图
The dimension of the galleries of
Guggenheim Museum, New York

毕尔巴鄂古根海姆美术馆展厅尺度示意图
The dimension of the galleries of Guggenheim Museum, Bilbao

Contemporary art often has an explosive scale, but the museum must consider where to use large scale and small scale. The scale can be as large as needed for visual impression and public functions, but for human and personal use the scale must be comfortable. But, how do we constructively solve the transition issues between large space and small space? Traditional Chinese cities set precedents for the construction of both large and small spaces and their interfaces. They give us examples that might help avoid constantly seeking formidably scaled structures.

- Residences: small-scale units
  Residences usually use comfortable small scale, which is not suitable for public spaces.
- Public spaces: the amplification of small-scale units
  In urban context, the use of large scale is decided by height and by extension laterally as well.
- Gardens: small-scale partition within a large scale
  Gardens and landscaped settings soften the boundaries between large and small spaces by integrating the interior and the exterior naturally, and thus reconciling successfully the differences in large and small scales.

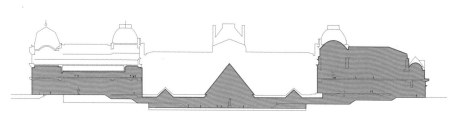

卢浮宫展厅尺度示意图
The dimension of the galleries of the Louvre

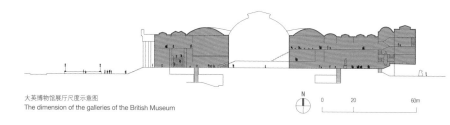

大英博物馆展厅尺度示意图
The dimension of the galleries of the British Museum

N
0   20   60m

住宅尺度：宜人、亲切的建筑空间
Residential Scale: humane and affinitive architectural space

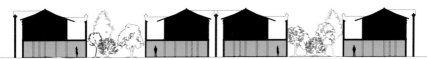

城市尺度：水平复制与组合
City Scale: horizontal duplication and combination

园林造境尺度：模糊柔和而又变化多端的边界
Garden and landscaped setting: softened, blurred and changeful boundaries

大和小的悖谬

常见的设计问题是把小尺度的空间等比放大，人的尺度和放大后的空间产生某种不对称的悖谬。基本画廊单元应有宜人的尺度。

The Confusion of Big and Small

One common design mistake is to copy and enlarge small-scale spaces in proportion, causing confusion of the human scale in the enlarged spaces. In general, galley spaces need to be set in human scale with comfortable dimensions.

大和小的结合

敦煌石窟可以看作由一系列小尺度"画廊"组成的大尺度"美术馆"。

The Combination of Big and Small

Dunhuang Caves can be viewed as a large-scale museum composed of a series of small-scale "galleries".

大和小的边界

如果比较故宫和圆明园，前者建筑众多使游人疲于奔走，后者的一系列小园子由于柔和的景观分隔，并不让人觉得有明显的界线和生硬的空间转化，因而感到放松。

The Border of Big and Small

The Forbidden City poses a great contrast to the Old Summer Palace. For most of its visitors, the former could be exhaustive for numerous buildings, while the latter makes the visitors more relaxed, as soft transition is made with landscape elements in a series of small-scale gardens, without obvious borders.

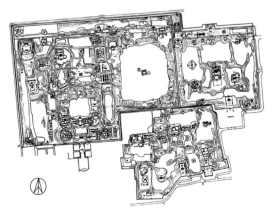

中国艺术和中国空间

Chinese Art and Chinese Space

中国空间
Chinese Space

是否存在一种本源性质的"中国空间"？中国传统生活的美学是否能和当代建筑无缝对接？理论家们对此持有不同看法。同样，当代的"中国艺术"早已不是铁板一块，美术馆设计大可不必为自己预设僵化的前提。

然而，现实中建筑设计及其评价又不能不受制于现阶段人们对"中国气质"的看法，由于当代中国文化愈发强烈的自我认同需要，这种看法变得日益重要。与此同时，当代中国美术馆展出的艺术品有部分还是传统样式的，在此，理应提供一些相对贴近中国艺术传统的建筑学主题。

Is there really an essential "Chinese space"? Can the aesthetic of traditional Chinese life be seamlessly stitched into contemporary architecture? Here theorists have different opinions. Moreover, contemporary Chinese art is not a unified whole. Therefore, museum designs do not necessarily restrict themselves with a rigid, predetermined, so-called rubrics.

However, in reality Chinese people's views of what constitutes "Chinese character" have inevitably affected the actual state of affairs in which people design and evaluate buildings. As a result of an increasing need a Chinese identity in the cultural perspective, this emphasis on "Chineseness" has become even more pronounced. Moreover, a large portion of exhibits displayed in contemporary Chinese museums are still traditional art. Therefore, it makes sense to provide some architectural topics related to Chinese art traditions.

内在的形态塑造
Center-Focused Form-making

中国传统注重内在的形态塑造，外方内圆，以不变应万变。迄今为止，它似乎没有那么关注构造的精确性问题，但对于细节却有着人性化的关注。虽然中国传统建筑不能和当代美术馆的功能要求相提并论，但是，通过强调内外造型的关系，为塑造具有中国气质、容易为公众接受的现代建筑形态提供了一种有益的思路。

Chinese tradition stresses center-focused form-making with human-oriented details instead of construction accuracy. Even though traditional architecture cannot stand up to contemporary architecture in terms of museum scale and function, the former provides an insightful approach for building a modern architectural form with Chinese characteristics and easily accepted by the public.

里和外的转换

江南园林的重要特色是由外部城市的规整形态由外而内过渡到人际的柔和感受。

The Transition In and Out

One important feature of Jiangnan Gardens is the gradual transition from the standardized form of the city into the soft and humane feel of the green space.

外方内圆 **园林建筑**

建筑外部轮廓方正简洁，内部空间丰富变化。在外部，建筑师需要防止刻意地追求和中国纪念性公共建筑形象不符的建筑造型；在内部，建筑师需要防止以室内被动地适应室外，在表皮内塞进难以使用的空间，或是以一成不变的方格子填入不同的建筑体量。

Round-Angular **Garden Space**

A building may enrich its orderly profiles and simple forms with dynamic interiors. For the appearance, the architect should avoid arbitrary building forms that are not compatible with Chinese monumentality. In addition, the building interior cannot simply be a passive "finish" that blindly follows the exterior form, with unusable spaces inserted into the building, or with immutable grids filling dissimilar building masses.

外宽内紧 **构造逻辑**

建筑的实际感受需要轻松，但内在的结构逻辑需要清晰。建筑师既要避免使用感觉过于冰冷的建筑材料和设计手法，也要明确控制内部空间的总体逻辑和建造方法。

Loose-Tight **Architectonics**

A building may appear soft and relaxing but its internal logic needs to be clear. A museum architect should avoid abuse of "cool" materials and design approaches, and should have a tight control of the overall logic of space-making and the architectonic approach to the building.

拐弯抹角 **构造细节**

建筑形体大轮廓清晰，不妨碍局部细节的到位和宜人。建筑师需要防止将简单呆板的小尺度形体简单放大，套用于体量空前巨大的公共美术馆项目。"大"和"小"应该适当结合。

Profile-Detail **Construction**

While a building may look neat in contour profile, its details can still be tangibly in place and pleasant. A museum architect should avoid applying the models for small-scale buildings to large-scale public museums. The "big" and the "small" should be suitably integrated.

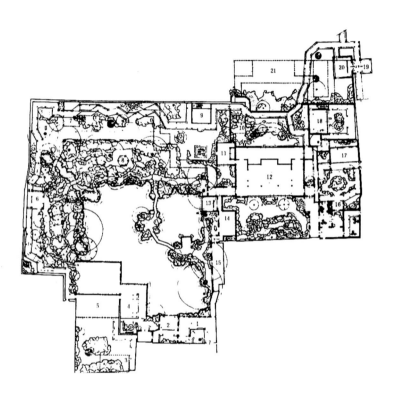

规整 vs. 有机

留园：规则的建筑格局中穿插着有机的景观元素。

Orderly vs. Organic

The Lingering Garden, Suzhou. Orderly architectural
layout is mingled with organic landscape elements.

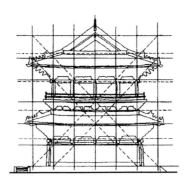

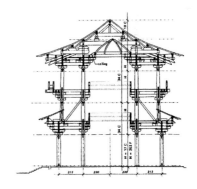

构造逻辑和细节

独乐寺观音阁的立面和剖面研究；虽然建筑外部轮廓转折多变，内部空间却是因地制宜。建筑物的总体结构遵循清晰的建造逻辑。

Architectonic Rationale and Detailing

Façade and sections, The Boddhisattva Pavilion, Dule Temple. Despite its angular silhouette, the pavilion's interior space follows the topography and retains a clear architectonic logic.

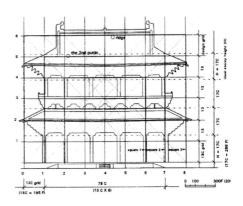

中国艺术中的色彩
Colors in Chinese Art

没有孤立的所谓"中国色彩",但中国人对特定色彩效果的喜爱有其"情境逻辑":中国文化倾向于动态地、整体地看待色彩效果,用多层次的、变化的光线烘托色彩表现,避免强烈的、直接的感官刺激,个别的色彩呼应于整体的视觉秩序。

- 中国传统绘画艺术中很少单独描绘阴影,在描绘空间时不追求戏剧性光影效果和怪诞的三维透视;
- 中国传统建筑中几乎见不到玻璃、镜子这类强烈反光的介质,浓重的色彩一般隐藏在阴影之中;
- 中国传统物质文化中有一些半透明的介质,如瓷器、琉璃、玉器等,它表现的是一种经过控制的,空间深度不确定的透明性。

There is no isolated or self-contained Chinese color palette. Rather, the Chinese preference for certain color schemes follows a "contextual logic": Chinese culture emphasizes a more dynamic interpretation of color scheme and effect as a whole. Color is set off by using layered, alternating light patterns, so that direct and strong sensations are avoided, with individual colors echoing the entire visual experience and pattern.

- In traditional Chinese drawings and paintings, there are very few occasions directly depicting shadows. Dramatic light and shadow effects and unusual 3-D perspective are rarely employed in depicting a space.
- Chinese traditional architecture is almost entirely devoid of glass and other highly reflective surfaces. Powerful and attention-grabbing colors are usually hidden within shaded and less-conspicuous areas.
- Semi-transparent materials such as porcelain, colored glaze, and jade in traditional Chinese culture present a controlled and reserved transparency.

第五章

看的议题
Ways of Seeing

○ "看"是美术馆中核心的议题。

"Viewing" is central to museum studies.

○ 只是"看"并不透明和简单。

Yet viewing is not transparent and simple.

○ "看"的方式随文化和历史而不同，艺术史是了解"看"的主要渠道。

"Viewing" varies with culture and history; art history is the major channel to understanding it.

○ "看"并不意味着"看到了"。

Looking does not necessarily mean seeing.

○ "看得清楚"也不等同于"看得明白"。

Visual clarity does not necessarily mean intelligibility.

○ 美术馆的物理构成对"看"有明确的引导作用。

A museum's building construction guides the way of viewing.

○ 建筑师决定了美术馆如何"看"，也请听听艺术怎么"说"。

Architects decide on what to "see" in a museum. Let's listen what art has to "say".

画廊和画框

Gallery and Picture Frame

画 廊

"画廊"（galleria）是一种由来已久的普遍建筑样式，它是解析美术馆样式的基本元素。图卢兹的中世纪学者吉多尼斯（Bernardus Guidonis）曾解释说，"画廊"其实就是普通的"通道或走廊"——将艺术品错置于"画廊"的柱间，有利于形成一种行进中的观感。古代罗马人的住宅中不乏这种建筑样式，只是他们的"画廊"的柱间放置的不一定是绘画。"画廊"的基本要素是：

- 墙和柱；
- 墙和柱围合的区域形成的"展厅"；
- 展台或展柜或其他展览家具；
- 室外光或室内光；

Gallery

The "galleria" is an ancient and common building type and an essential component in analyzing the museum. Bernardus Guidonis, the Medieval scholar from Toulouse, explains that "galleria" is just a common corridor or hallway; by placing the artwork among the colonnades of the corridor, it produces the spectacle of a march or parade. The galleria is commonly seen in ancient Roman residences, but not necessarily adorned with paintings. The essentials of galleries are:

- Walls and columns;
- Display space, formed by enclosures made of walls and columns;
- Showcase, display counter, and other exhibition furniture;
- Daylight or artificial light.

北京中间美术馆，沿行进方向在墙面上布置2D艺术品。
Inside-out Art Museum, Beijing. Two-dimensional artworks are placed on walls along the direction of walking.

美国肯恩大学画廊，有室外光。
Kean University, New Jersey, USA. A gallery with natural light.

大英博物馆瓷器馆，典型的展柜和行进方向的关系。
Porcelain Gallery of the British Museum. A view of showcases and direction.

□美术馆三号馆旧建筑形成的跃层空间将上下两个不同的画廊连接在一起。
day Art Museum, No. 3 Gallery. A mezzanine space converted from an old building connects two galleries at different floors.

美国亚利桑那州立大学博物馆陶瓷艺术展厅，展示3D展品的展厅有室外光。
Arizona State University Museum, ceramic art gallery. The gallery, displaying 3D artworks, introduces natural light.

画框

"画框"既是文艺复兴以来西方绘画重要的物理组成,也是一种具体而常见的观看方式。它促成了一种透视空间中的幻觉性再现,其意义不仅限于绘画作品。"画廊"的柱间建筑元素本身就可以构成"画框",美术馆室内室外空间的各种垂直和水平构件相当于"画框"。中国艺术中的"框景"和"画框"的含义有所重叠但绝不相同。

对照罗马人放置雕像的柱廊,或是与文艺复兴时期画家创作出来的透视空间与建筑细节合而为一的情形相比,现代"画廊"中的情形要复杂得多:首先人们观看绘画作品的方向和行进的方向是不一致的,同三维的雕像不同,平面的绘画通常只有在正面观看才是富有意义的,其次"动"(沿着画廊走动)"静"(停下来侧身观看)的两种行为模式也难相容。

现代美术馆发展的大势是"行动"和"观看"发生了一定的分离。

Frame

The frame has been both a physical component of Western painting since the Renaissance and a very common way of viewing. It promotes an illusionistic representation in perspective and extends beyond the two-dimensional works. Architectonic elements in a galleria actually make "frames". A variety of horizontal and vertical components seen in the museum space can also act as "frames". Chinese art too contains framed scenery or frames, with overlapping but not equal connotations.

Romans installed the sculptures between arcade colonnades. Renaissance painters created an illusionistic space in which two-dimensional works extended architectural visions. The case of modern galleries, by contrast, is more complicated. First, the procession of the viewer along the wall poses a directional difference from the viewing of paintings facing the wall. Vision is the best for 2-D paintings when the viewer faces the wall, unlike the vision from all directions for 3-D sculptures. Second, this processional aspect of moving sets itself in opposition to the stillness of viewing.

The most recent development of the museum is about a separation between two behaviors: moving and viewing.

1. 没有"框内"只有"画框":艺术家肖
   昱对"画框"的戏思。
   No "content" but only "frame": parody
   by the artist Xiao Yu on "framing".

2. 室外悬挂的画框定义了加尔各答的街头
   美术馆。
   Outdoor frames define a street
   museum in Calcutta.

3. 杭州三潭映月景区桥下。
   A natural frame: under the bridge of
   Santan Yingyue, Hangzhou.

4. 俯瞰佛罗伦萨的"画框"。
   Frame overlooking Florence, Italy.

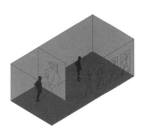

矛盾的画廊（既是"背景"又是实体空间）
老迈腾斯的肖像画常常以一个画廊作为背景。
这里的画廊既是一个实有的空间，又是一种
身份和地位的象征。

A Paradoxical Gallery (as background
and space)
Portraits by Daniel Mytens the Elder often
take a gallery as backdrop which is both
a real space and a symbol of the figure's
social status.

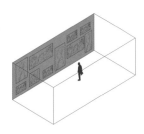

文艺复兴初期画廊样式（大量画框的聚合）
A Gallery in Early Renaissance
(Juxtaposition of multiple frames)

画廊既可以看作一个实有的空间，也可能是物质世界向抽象意义演进的"转换器"。

The gallery can be viewed as both a solid space and a "transformer" of the physical world into abstraction.

随着现代意义的美术馆的出现，空间和形象发生了分离。也就是说，作为体验场所的环境和作为图像产生机制的幻觉性表面不再一定相关了。当代美术馆将两者重新定义，又用创造性的方式组合在一起。比如有的美术馆偏重于公共活动，变得更像一个热闹的社交场所；而另外一部分美术馆则加强了"观看"的属性，美术馆成了剧院。

With the emergence of the museum in the modern sense, space and image manifest themselves distinctly. That is to say, the museum as a locale for experience is not necessarily related to its illusionistic character as an image-generating mechanism. The contemporary museum redefines them both while again integrating them in novel ways. Some museums are heavily loaded with public events, transforming themselves into busy social spaces while other museums are dominated with "viewing", turning into enormous theaters.

罗马人的雕像展廊（实体空间和画框并存）
Roman's Sculpture Gallery
(Juxtaposition of physical space and
"frame")

全景画（环境性观看）
Panorama (Environmental viewing)

重屏会棋图（前后重叠的画框）
The "Double Screen" (Overlaid frames)

云岗石窟（晦暗空间中的观看）
Yungang Grottoes
(Viewing in dimly-lit space)

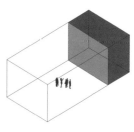

巴洛克剧院（宏大单向的画框）
Baroque Theatre
(Grand single-way frame)

幻觉性透视空间（实体空间中的画框消失了）
Illusion in Perspective
(Disappearing frame in physical space)

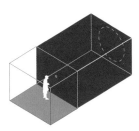

窥视（以小观大）
Peeping Holes
(The big seen in the small)

假 3D 空间（平面图形模拟出立体的效果）
Pseudo 3D space
(3D imitated with 2D)

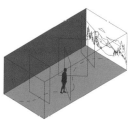

"如画"（自由探索的多视点"游赏"）
Picturesque (Multi-point tour for free exploration)

行进视廊
（城市作为一种无尽展开的展场）
Processional Corridor for Viewing (City as an endless gallery)

一般而言建筑物只是容器，它的功能是预设的，不会轻易改变，而且和它的内容也没有太大的关系。但是一座美术馆的核心功能很大程度上却取决于它的收藏，以及人们对于收藏的不同"看"法。这是一座美术馆应有的使命。

因此设计美术馆势必要对艺术史有所了解，在深入理解艺术发展的基本规律的同时提出创新的概念。一座美术馆的创新概念可能不符合现在的功能要求，但是却有可能为将来很长时间的艺术和社会生活发展开创新的方向，这是历史上不乏其例的。

对于承前启后的中国当代艺术而言，尚有很多未定的东西需要探索，美术馆设计既要关注建筑的"形式"，也要关注其独特的"内容"。

In general, buildings are only containers. Their functions are preset and difficult to alter, not necessarily related in any organic way to their contents. On the contrary, the core function of the museum is determined to a large extent by its collection as well as by how spectators view the collection. This is the very raison d'etre of the museum.

Therefore, it is necessary to know art history in order to design good art museums, and to innovate on the basis of a deep understanding of the historical development of art. Innovative design of a new museum might not fit the current "container" functions but may set an entirely new direction for the development of art and social life in the long term. It's not difficult to find such examples in history.

For Chinese contemporary art, which connects the past with the future, there remain many unknowns to explore. Museum design needs to be attentive to both building type and the unique content of its institution.

观望和凝视
Watch and Gaze

由于展示理念的不同导致了不同的建筑布置，同时这些布置又反过来形成了对于展品的新的理解。

通常的美术馆布展方式是只让一幅作品占据墙面（或一个观看方向）的中心。这样单一的视角往往和艺术品至高无上的地位联系在一起，它有强烈的"画框"属性，往往产生一种戏剧性的舞台效果，在其中只有一个方向的观看才有意义，这种观看带来局部聚焦的清晰度和逐渐深入的心理暗示。

在一些画廊中，建筑利用内部装饰形成了展品和空间一体的情态：例如，在德国波茨坦的无忧宫中，由纵贯檐部的三段式柱式形成了壁龛形的空间聚合，在其间可以放置艺术品，并非突出单一的"作品"。这些建筑元素实际上起到了隐形"画框"的部分作用，它使得观众的观看形成某些确定的段落，但并不撕裂环境和作品。

Different exhibition models lead to vastly dissimilar architectural settings. Conversely, the settings contextualize new understandings of what is displayed.

Museums commonly display a single work as the only object of the wall (or of the vista), securing the primacy of the art work. The result is a highly visible frame and a theatrical experience in which only a single way of viewing manifests itself. The viewing, then, orients itself to a singular visual focus, psychologically drawing the viewer deeper into the pictorial space.

In some galleries, the museum unifies the art objects with interior spatial decorations, giving blurred boundaries between art and its (con)text. For example, in Sanssouci Palace of Potsdam, Germany, the Classical orders forms a type of a niche-like space where art-objects can be displayed. These architectonic elements do not emphasize the singularity of the works, but instead contextualize them through invisible frames. They punctuate the viewing experience of the spectators but do not completely separate the work from its surrounding context.

2011 西安园艺博览会一景
A Scene at Xi'an Horticultural Exposition 2011

大都会艺术博物馆的中国古代艺术展厅
Gallery of Ancient Chinese Arts at the Metropolitan Museum of Art

环境性的观察
Environmental Observation

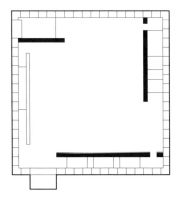

布雷根茨美术馆首层平面图
Plan of the first floor, Kunsthaus Bregenz

布雷根茨美术馆典型展览区平面图
Plan of typical gallery floor, Kunsthaus Bregenz

与此同时，初期的"画廊"也有相当多将画一幅幅密布在墙面上，使得它们几乎形成一个完全连续的表面，这样的做法也许和那时许多"画廊"也是绘画交易场所的功能密切相关，无论如何其结果是第一种情况的反面：这些展览场所中，空间的"段落"和节奏变得模糊，"凝视"的局部深度让位给环境性的静态感受了。

At the same time, a large portion of early galleries had a large number of works densely hung on the wall side by side, almost forming a single art-surface – a common practice possibly related to the function of the gallery as an auction place. Nevertheless, the result is the opposite to the single-work display model. In such galleries, the punctuation and rhythm of spectacle are blurred. The depth of various vocal points gives way to a static experience of the environment.

右页图
彼得·卒姆托设计的布雷根茨美术馆。建成之后的美术馆经由不同的展览呈现出多样化的空间意图。同一个建筑带来广阔多样的"看"的舞台，其中美术馆自身往往趋于消失。

Right
At the Kunsthaus Bregenz, Austria, designed by Peter Zumthor, the built museum presents a diversified spatial experience with a variety of different exhibitions. The same building brings about a spectrum of stages of viewing possibilities, in which the museum itself disappears.

聚焦和错视
Focus and Illusion

"看见了"并不意味"看到了"；"看得清楚"也不意味"看得明白"。
事实上，很多艺术作品有意识地利用了"错视"和"幻觉"。

如果我们将空间比喻成一个照相机的话，"错视"就是一张"失焦"的相
片，"幻觉"则可能是两张错误地重叠在一起的照片。在美术馆设计的情
形里，这种"故障"也许反而是个亮点。

德国文艺复兴艺术家阿尔费雷德·丢勒成像机制的装置
Illustration of image-making by Albrecht Dürer, an artist in
the Renaissance period

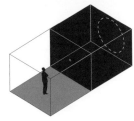

"照相机"强调的是光学记录的客观特征。但"照相机"并不是在空间中观看的唯一模型，当观看者和被观看的东西没有固定的关系的时候，另一种类型的"看"的空间就出现了。

A "Camera" emphasizes the subjectivity of optical representation, but the camera is not the only means by which one can see the space. When the viewer and what is viewed do not have a fixed relationship, there emerges a new type of possibility to "see".

Looking does not necessarily mean seeing. Visual clarity does not necessarily mean intelligibility. In fact, many artworks purposely take advantage of illusion and visual fallacy. If we liken the space to a camera, the visual fallacy leads to an "out-of-focus" picture while the illusion might be two photographs mistakenly overlapped. In the case of museum design, these "errors" might become the museum's highlight.

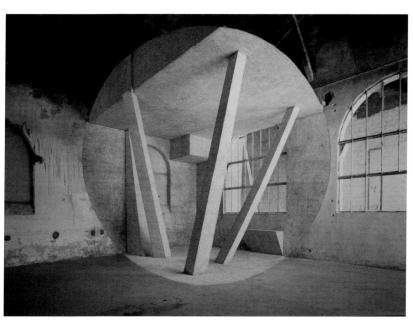

乔治斯·胡塞的摄影作品呈现出多个重叠在一起的空间。
Georges Rousse's photography work presents the space in overlapped visions.

中国艺术传统的影响
The Influence of Chinese Art Tradition

....

不同之"看"

Different Viewing

不太合乎西方标准的中国艺术品样式让"不标准"的美术馆空间变得可能。

● 展览空间是否一定要"四白落地"一览无余?

● 展览空间是否可以变得更个人化?

● 展览空间是否可以探索视觉之外的可能?

● 展览空间是否可以变得非物质化(数字化……)?

Chinese art that is not yet completely Westernized makes possible a "non-standard" museum space.

● Are exhibition spaces supposed to be all "white" with unobstructed views?

● Can exhibition spaces be more personalized?

● Is it possible for exhibition spaces to explore senses beyond vision?

● Could exhibition spaces be immaterialized (digitalized…)?

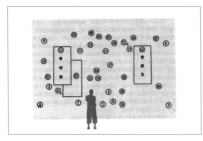

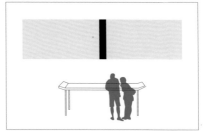

叙事的空间：敦煌降魔变

观众的眼光需要在壁画不同的部分之间穿梭往来，整个画面也是先后有序地按照情节来组合。

Space of Narration (Transformation Tableaux of the Subjugation of Mara at Dunhuang Caves)

The pictorial program of the painting is organized by the order of the plot. Visitors need to move their line of sight to different scenarios contained within the same mural composition.

连"环"画·五代：顾闳中《韩熙载夜宴图》手卷（局部）

《韩熙载夜宴图》手卷的各部分之间形成段落。同时又彼此呼应。现代博物馆将手卷展开，一览无余的展示多少破坏了这种讲故事的模式。

A Painting Gradually Unfolded (details of Han Xizai's Banquet, by Gu Hongzhong, Five Dynasties)

Sections of the scroll painting make punctuated "paragraphs" for the viewer, suggesting possible connections and correspondence. Modern museums display the scroll painting unfolded as ONE long picture, disrupting the supposed pictorial program of the work which unfolds the story for viewers section by section.

很多人认为，从展品的艺术史含义倒推空间设计是展览设计正确的工作方向，但是在现行体制下，很多按照标准模式设计的建筑空间一旦落成，"追加设计"已经不能改变大局，实际上已经排除了这些创新型展示方式结合个别艺术品特点的可能：

● 展览空间是否可以同时再现"物"和"像"？

● 展览是否可以把展览空间自身作为展览对象？

● 展览空间是否可以同时展现艺术品的物质载体和它的抽象意义？

● 同一展览空间是否可以展示多个艺术家作品？

被隐藏的图像：中山王陵《兆域图》

被深藏于陵墓及秘府中的古代平面图似乎不是为了给生人观看的。

非图像的艺术：唐·张旭《肚痛帖》

书法至此已经不仅仅是一种表意符号，但它也不是一幅画，它是一种抽象的图像。

Hidden Image (Zhao Yu Map from the Tomb of King of Zhongshan)

The ancient map buried deep in the tomb was seemingly not meant to be seen by living people.

Non-Figurative Art (Calligraphy by Zhang Xu, Tang Dynasty)

Calligraphy goes beyond a symbol. Neither is it a painting. It is an abstract picture.

People might think that spatial design based on art history falls into the correct category of display design. But standardized museum spaces, once built, in fact prevent themselves from being remedied or adapted to anything significantly incompatible, which excludes the possibility of making innovative and idiosyncratic exhibition displays.

- Can exhibition spaces present both the object and its image?
- Can an exhibition space treat itself as an object to view?
- Can an exhibition space present the materialized form of an art object as well as its abstract meaning?
- Can collective works of artists be displayed in one exhibition space?

拓片：战国水陆攻战纹铜壶

按照西方美术馆的标准，拓片是某种印刷品或复制品，属于"复数的艺术"。但一幅拓片未必仅等于原作的一幅影像，在中国艺术史中它或许具有和原作同样的价值，并且记录了从原石／金到拓片的人事和历史。

Rubbings (Bronze vessel from Warring States Period on Land and River Battles)

By the Western standard, rubbings are prints or copies of certain categories, a type of art in collective form. But a rubbing can go beyond a mirror image of the original. In Chinese art history, rubbings might carry the same value as the original, as they record the history starting from the original and ending with people who made rubbings.

壁画：莫高窟·盛唐第 45 窟

"装饰"与具象并存，平面的图案结合立体的彩绘，按西方美术馆标准，人们并不容易找到视觉的"焦点"与"画框"。

Mural (Dunhuang Cave #45, Tang Dynasty)

Figures in mural paintings coexist with decorations. Two dimensional patterns go with three-dimensional color graphics. It is not easy for viewers to find "focus" and "frame", by Western standard.

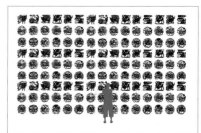

作为过程的艺术：唐西州习字残片

对"原作"的模仿和演练也可以成为艺术展示的一部分。

剪纸：民间艺术

目前权且以"民间艺术"命名的这类艺术品的标识就是"大量""集体"。

Process as Art (Pieces from calligraphy practice in Xizhou, Tang Dynasty)

The mockups and practices based on the original may become part of the art exhibitions.

Paper Cuttings (Folk art collections)

Temporarily termed as "folk art", this type of artworks is prominent for its "quantities" and collective quality.

中国艺术史中的一些实例将改写典型西方美术馆展览中
"看"的标准含义：

软	中国传统绘画不像发源于欧洲的油画，没有一个绝对平坦的绘画平面，没有意味强烈、指示明确的"画框"，也不一定要挂在墙上展示；
小	某些中国艺术品的尺寸相当小，不适合远距离观看；
表演性	某些中国艺术品的展示常和它的创作结合在一起，从而带来表演性的氛围，"看"不再是静态的；
灯光	某些中国艺术品除了需要人工光源照明，亦需要环境照明以营造特殊气氛。

Cases from Chinese art history will redefine the meanings of "viewing" in typical Western museums.

Soft	Chinese traditional paintings differ from European paintings in that they have neither a definitely flat surface of medium, nor a designated "frame" with specific meanings, and do not necessarily have to be hung on the wall for display.
Small	Certain Chinese art works in quite small dimensions are not suitable for viewing at a distance.
Performative	Certain Chinese art works integrate their display with their creative process, bringing a level of theatricality to the viewers, making "viewing" no longer static.
Lighting	Certain Chinese art works require not only artificial spot lighting, but also environmental lighting as well to create a special exhibition atmosphere.

1   美术馆＝雅集空间
     谢环《杏园雅集图》（大都会艺术博物馆藏本）

     Museum = A Gathering Site
     Elegant Gathering at the Apricot Garden,
     Xie Huan, The Metropolitan Museum of Art

2   美术馆＝创作空间
     中国传统书画的典型欣赏环境

     Museum = Art Studio
     Typical scenario for appreciating traditional Chinese artworks

Mounting

古人云："三分画，七分裱。"装裱艺术不仅是保护和装点传统中国书画的一种技术手段，它还意味着不同于西方画廊模式的陈设和欣赏模式。例如手卷通常需要一部分一部分地打开观赏。

A traditional Chinese saying notes that, "an artwork of painting is made of 30% painting and 70% mounting." The traditional mounting techniques of Chinese painting are not only a means to protect and adorn artworks, but also a suggestion of an Eastern model for artwork display and appreciation, which is different from the Western gallery model. For instance, a scroll painting is supposed to be unrolled and viewed section by section.

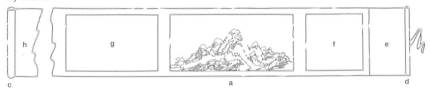

谢柏柯《中国绘画艺术》中所示中国传统艺术样式的物理形态

Traditional Chinese Painting Formats as Summarized in Chinese Painting and Calligraphy Style by Jerome Silbergeld

a. 手卷
   Handscroll

b. 立轴
   Hanging scroll

c. 手卷轴头／立轴地杆
   Handscroll's roller/Hanging foot roller

d. 手卷轴头／立轴地杆
   Handscroll's roller/Hanging head roller

e. 天头
   Inscription panel

f. 引首
   Title sheet

g. 跋纸
   Colophon panel

h. 拖尾
   End roll

i. 双面册页
   Double-leaf album painting

j. 单面册页
   Paired single-leaf album paintings

k. 单面册页的"蝴蝶"装
   Paired single-leaf album paintings, "butterfly" mounting

l. 团扇
   Screen fan

m. 折扇
   Folding fan

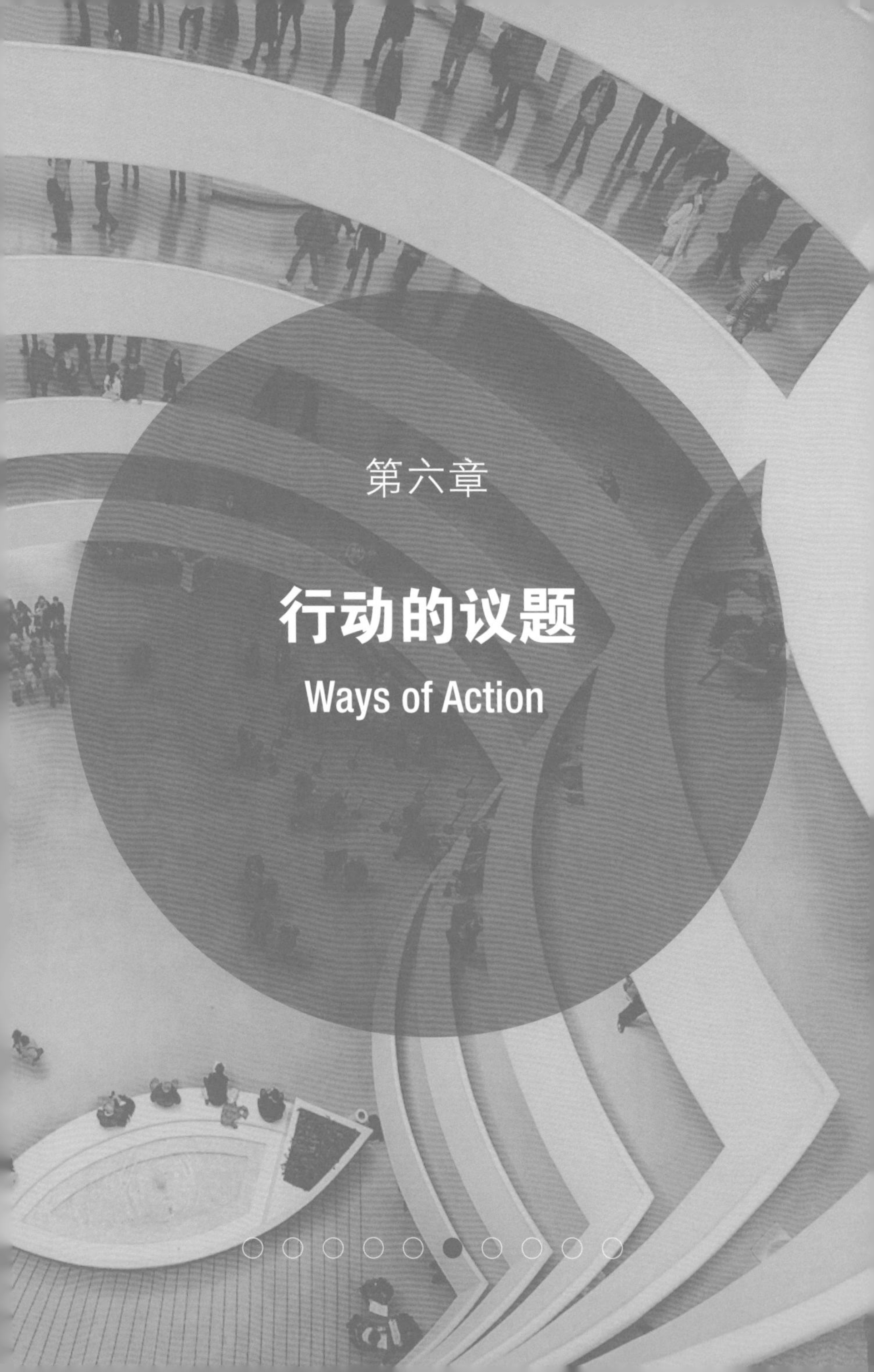

第六章

行动的议题

Ways of Action

○ 现代社会的条件决定了美术馆再难静观。
The nature of modern society means that museums are not quiet anymore.

○ 展览的体验却在动静之间。
Exhibition experiences swing between the still and the dynamic.

○ 流线图可以帮助我们理解美术馆中的行动。
Routes of circulation are the map to understanding actions.

○ 但展览并不是美术馆中公共活动的全部。
But exhibitions are not the only events held in the museum.

○ 设计"活动"，关键在于是否需要统分和控制不同的"运动"。
The key to devising museum "events" is whether to divide or control different "activities".

○ 公共空间的"活动"，永远有着开放的可能。
"Actions" held in public space are forever open.

○ 最自然的行动模式是依靠直觉的。
The most natural action model is intuitional.

○ 美术馆的导览应该遵循简单的原则。
Museum orientation should be simple.

动静之间的美术馆
A Museum Still and Dynamic

辛克尔设计的德国国家美术馆老馆最初面临着德国国家的新的城市语境。辛克尔的设计图凸显了美术馆面向城市的意图。这种打破"内""外"的做法在当时堪称创举。

The Altes Museum designed by Karl Friedrich Schinkel found itself in the new urban context of the modern German State. The design highlighted Schinkel's intention to keep the museum open to the city, an unprecedented practice to connect "inside" with "outside".

美术馆中"行动"问题的关键在于：我们是否需要控制公共空间使用者活动的种类和规模？对艺术的理解和思考本需要个安静的氛围。德国浪漫主义者威廉·海因里希·瓦肯罗德表示，"画廊应该是庙堂……艺术本质上不属于寻常生活之流，属于神的思绪。"早期美术馆的创立者，比如国家老馆的奠基人之一、伟大的人文主义者洪堡认为，人们参观美术馆的次序和内容都要严格地控制；然而，现代社会越来越不鼓励这种自上而下的"控制"。在每个使用者都具备自己的意志的前提下，作为一类重要公共空间的美术馆再难"安静"。

The key consideration to understand the "actions" of museums is: Do we need to tightly control the categories and scales of events in public space? To understand art and reflect on it, it demands a quiet environment. The German Romantic Wilhelm H. Wackenroder said, "Galleries ought to be temples [...] Art does not belong to common life but to the thoughts of gods." The humanist Alexander von Humboldt, one of the founders of early German museums such as the Altes Museum, argued that the museum must assume tight control over the order and content for museum visitors. However, modern society has come to discourage this type of top-down control. When each visitor has his or her own will, museums as important public spaces can no longer be "quiet".

隋建国作品装置现场。铁球会在展场中来回滚动,它打破了美术馆的静态。
Installation by Sui Jianguo. The steel ball rolls back and forth, breaking the stillness of the museum.

流线类型
Circulation Types

串联 / 并联 / 放射 / 自由 / 螺旋

传统的美术馆建筑流线型基于这样的假设：要么是强调时间的历时性"浏览"，要么是强调空间的并时性"博览"。不同的空间和时间是彼此分离、各自清晰的。在后来的美术馆中出现了允许"信步"的蓬皮杜国家文化艺术中心，以及允许"游目"的古根海姆美术馆。时、空在此混融了。

Linear / Annexation / Star / Free / Spiral

The traditional circulation routes of museums are based on the following hypotheses: It is either a chronological "browse" of discrete items, or a spatially organized "survey" of various juxtaposed art-objects. In this way, time and space are separate and distinct from one another. The modern development of museums has seen new types of museums, such as Le Centre Pompidou, which allows patrons to freely "roam" within the space, and the Guggenheim Museum in New York, where vision itself "wanders" without regard for an externally imposed itinerary. Thus, time and space meet each other.

1　**串联式**
参观者严格按顺序走过所有展厅。

2　**并联式**
参观者严格按顺序参观，但可以略过其中一部分展厅。

3　**放射式**
放射式流线类型和简单的并联式区别在于并列的各展厅之间并无前后次序的分别。

Linear
Visitors go through all galleries following a strict route.

Annexation
Visitors go through all galleries in a strict route, but they may skip some galleries on the way.

Star
The star type is different from a simple annexation in that the former does not have preset viewing sequence.

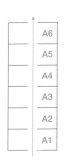

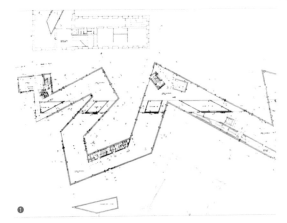

❶

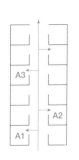

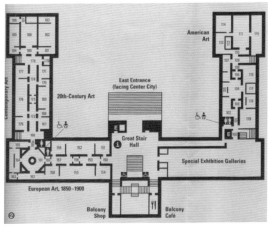

❷

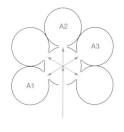

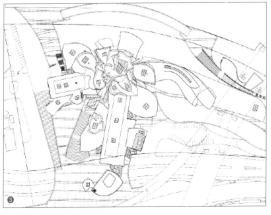

❸

不同流线自然形成的对美术空间的不同体验

即使在古典主义时期，博物馆的设计也常常整合了多种流线类型。例如辛克尔著名的国家老馆便在直线型的柱廊平面中嵌入了圆厅，使得对展览流线的体验有了多种可能。

Different circulation routes leading to different ways of experiencing museums

Even in the time of Neo-Classicism, museum design often integrated multiple circulation types. For instance, Karl Friedrich Schinkel combined a rotunda with linear galleries in the famous Altes Museum, offering multiple viewing route choices for visitors to experience.

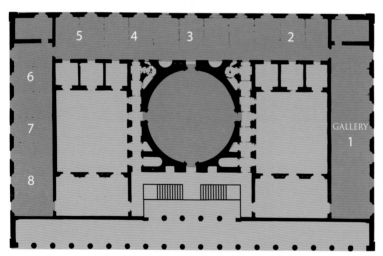

柏林老博物馆平面图
Plan of Altes Museum, Berlin

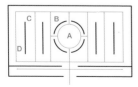

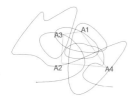

4　自由式

自由式流线类型是 20 世纪美术馆设计中出现的创新空间类型。它重新定义了看和行动的关系，也就是时间和空间的关系。在可以"信步"的自由式中，参观者并无时间的先后次序的拘束，因此它是在同一空间中混融了不同的时间。

Free

The free-style circulation route emerged in innovative types of 20th-century museum spaces. It redefines viewing and action's interaction in a new way, in response to new notions of the relationship between time and space. In the free-style "roaming" circulation route, the visitor is not bound to a time sequence, and thus may experience different moments within the same space.

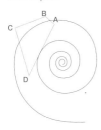

5　螺旋式

在可以"游目"的螺旋式中，参观者可以在一个确定的时间点看到不同的展品，因此它是在同一时间里体验了不同的空间。

Spiral

In the spiral circulation route where the vision wanders, the visitor can see different portions of the exhibition at a fixed moment and thus participates in multiple spaces at the same time.

德国柏林新国家画廊
Neue National Gallery, Berlin, Germany

纽约古根海姆美术馆
Guggenheim Museum New York

通达性
Accessibility

美术馆设计中的通达性指的是"可达"和"有方向感"。应当灵活、恰当地处理展览、公共空间和工作单元的通达性，并协调三者之间的关系。

展览、公共空间和工作单元各自的通达性要求是不一样的。

一般说来：展览应避免过长的、一成不变的、枯燥和没有间断的展线。各段展线间应穿插有趣的空间变化。公共空间应该具有相对更多的和城市对接的灵活性，使得参观者有选择的余地，并和展览项目有机融合，尽量减少步行距离。根据不同规模和性质的美术馆要求，办公室的管理者可以决定什么样的通达性要求更适合自己。以上三者的适当组合构成了一个美术馆空间设计的重要"输入"条件，也方便了将来的使用和管理。

Accessibility issues in museum design involve both accessibility and orientation. We should assess properly the accessibility issues of exhibition space, public space and office space and coordinate them in a flexible manner.

Demands for accessibility and orientation are different for exhibition space, public space, and office space.

In general, exhibition space should avoid long and monotonous routes in accessibility, which can be tedious and exhausting. Sections of circulation routes should be punctuated with interesting spatial intervals. Public space should offer flexibility in connection to the city and to exhibition space, giving visitors more choices with accessing the exhibitions and reducing walking distance. For office space, depending on the scale and the function of the museum, museum management can choose an accessibility model suitable for their needs. Combinations of the above three types of spaces in museums set an important "input" condition to space design in a museum and will facilitate its future use and management.

对于不同的使用者来说美术馆的空间是不一样的。
Different users experience the museum space in different ways.

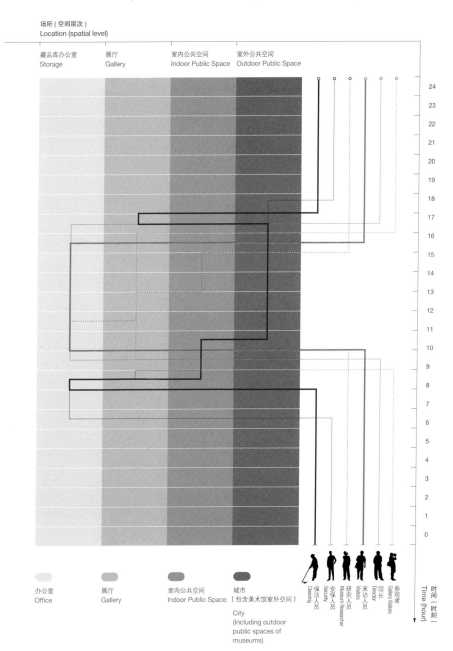

场所（空间层次）
Location (spatial level)

藏品库办公室　展厅　　　室内公共空间　　室外公共空间
Storage　　　　Gallery　　Indoor Public Space　Outdoor Public Space

24
23
22
21
20
19
18
17
16
15
14
13
12
11
10
9
8
7
6
5
4
3
2
1
0

办公室　　　展厅　　　　室内公共空间　　　　城市
Office　　　Gallery　　　Indoor Public Space　（包含美术馆室外空间）

City
(Including outdoor
public spaces of
museums)

保洁人员 Cleaning
安保人员 Security
研究人员 Museum Researcher
来访人员 Visitors
馆长 Director
参观者 Gallery Visitors

时间（时刻）
Time (hour)

公共空间和公共服务设计
The Design of Public Space and Public Service

与美术馆相交的城市空间（包含美术馆室外空间）、美术馆室内公共空间、展厅（展览空间）、办公（专业空间），将这四个对外对内的层次适配的主要功能交错搭接起来，就得到对美术馆功能进行新的定义，可针对不同使用者来设计美术馆空间形态。

Museum spatial functions can be roughly divided into four levels, ranging from the most private to the most public: Office (service areas), Gallery (exhibition space), Indoor Public Space, City Space that interfaces with the museum (including outdoor public space of the museum). The four levels of spatial functions can be intertwined to form a new definition for the museum functionality and create a different spatial configuration targeted at different users.

1–3　不同类型的美术馆空间（1. 大都会艺术博物馆中国花园／2. 大英博物馆展厅／3. 史密森尼美国艺术博物馆中庭）

Museum Spaces of Different Characters (1. Chinese Garden, The Metropolitan Museum of Art, New York / 2. Exhibition hall, The British Museum, London / 3. Atrium in Smithsonian American Art Museum, Washington D.C.)

公共空间的"活动"类型永远有着开放的可能，传统美术馆的设计意在控制与限制使用者的活动，新型美术馆则试图释放活动的可能性。比如，展厅能否直接面向城市？研究人员是否可以在展厅中办公？

Activities in public spaces offer many different options. Traditional museums tend to emphasize control and limitation over users' activities, while new museums tend to be open to different possibilities. For example, is it possible to keep the gallery space open to the city? Can researchers work in the gallery space?

办公
Office

展厅
Gallery

室内公共空间
Indoor Public Space

城市（包含美术馆室外空间）
City (including outdoor public space of museums)

来访方向
Visitors come this way

安保控制
Security control

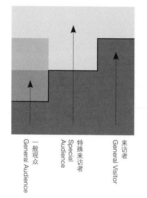

一般观众
General Audience

特殊来访者
Special Audience

来访者
General Visitor

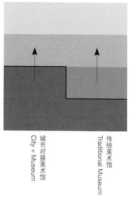

城市对接美术馆
City + Museum

传统美术馆
Traditional Museum

城市 + 办公

来访者直接进入美术馆办公空间，美术馆的庞大员工群、专业研究人员和城市空间可以直接对接。

City + Office

Visitors access the office area directly, which links the staff, researchers with the urban space.

城市 + 室内公共空间

城市到访者将这部分美术馆空间作为公共空间使用，对他们而言的方便，不一定涉及美术馆的专业展览功能。

Urban Space + Indoor Public Space

Visitors use this part of the museum primarily as a public space, which offers convenience without interfacing with the professional function of exhibitions.

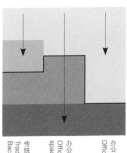

传统后台办公模式
Traditional Office Backstage

办公对接公共空间
Offices open to public space

办公对接城市
Offices open to the city

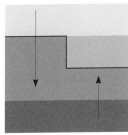

公共空间展厅合一
Public space as Gallery

城市对接美术馆
The museum open to the city

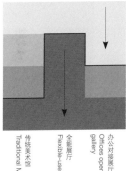

传统美术馆
Traditional Museum

全能展厅
Flexible-use gallery

办公对接展厅
Offices open to the gallery

室内公共空间 + 办公

对于美术馆的庞大员工群和专业研究人员而言，美术馆的室内公共空间应具有小型城市的基本交流功能和生活便利。

室内公共空间 + 展厅

美术馆的专业展览功能让室内公共空间的品质非凡，反过来，具有小型城市功能的公共空间把美术馆的专业使用变得更人性化。

城市 + 展厅

美术馆可以考虑将其专业功能由室内转向室外，由封闭空间转向城市开放空间。

Indoor Public Space + Office

For museum staff and management staff, indoor public space of a museum should serve as a mini city, which provides convenience for basic communication and daily life.

Indoor Public Space + Gallery

The museum's professional functions related to exhibitions can bring a unique touch to the indoor public space. At the same time, the indoor public space should serve as a mini city, providing professional uses with a more humane environment.

Urban Space + Gallery

A museum can consider the transition of its professional functions from the inside to the outside, from an enclosed space to the more open urban space.

Simple Orientation

对于一般观众而言，最自然和直觉的通行模式依然是"顺路"的参观和导览方式。巧妙的"顺路"需要创新的空间概念。

For average viewers, the most natural and intuitive way to see an exhibition is following a simple orientation without branching out. Innovative concepts of spatial orientation can revolutionize this approach.

老布鲁盖尔的《通天塔》
螺旋坡道
The Tower of Babel, by Pieter Bruegel the Elder
Spiral pathway

莱特晚年创造的古根海姆美术馆的螺旋形画廊空间引起了很大的争议。它打破了传统建筑"层"的概念，创造了一个自我复制但并不重复的混融空间。人们在这种美术馆中绝无迷路的可能，同时也别无选择。但它又不同于传统一条道走到黑的美术馆，因为在这里人可以对即将经历的空间"提前"有一个整体的、几乎是即时的把握。

The spiral gallery in Guggenheim Museum designed by Frank Lloyd Wright was controversial. It breaks the concept of floors in traditional buildings, by creating a self-replicating yet non-repetitive space of fusion. Visitors can never get lost, with no alternative choices for routes. But the spiral pathway is not a simple linear route, as people almost immediately have expectations of the whole structure which they are to experience.

洛杉矶郡立美术馆日本艺术展亭，布鲁斯·高夫设计
室内及室外的螺旋坡道
The Pavilion for Japanese Art, Los Angeles County
Museum of Art, designed by Bruce Goff
Spiral pathway indoor and outdoor

简单的原则
The Principle of Simplicity

公共空间和公共服务的设计，应为观众和工作人员提供一种新颖有趣、层次清楚并易于理解的总体逻辑，便于使用者在巨大的尺度中导向和快速到达。公共空间和公共服务的通达性体现在使用者可以方便定位、即刻理解，例如通过色彩识别不同的功能定位。

公共空间和公共服务设计应从真实的城市生活中得到灵感，它的主要原则在于空间的使用效率。首先，使用者有多样化的选择，例如在同一地点使用者可以去往不同的餐饮地点；同时，同一公共空间可以承载着多样化的功能，例如使用者在去往展厅的路上可以顺便光顾艺术书店，艺术书店同时可以提供简单的餐饮，等等。

The design of public space and public service needs to provide an innovative and easy to understand overall logic for both visitors and staff. Both groups of users should find it easy to navigate and get oriented within the large space in order to reach their destination quickly. Easy access to public space and public service means convenient positioning and instant understanding, for instance, the use of a color scheme to help to identify different function areas.

Museum architects should draw inspirations from the daily urban life, and focus on using space efficiently. First of all, users should be given a wide range of choices, for example, from one area to different dining points. Second, one public space can be equipped with multiple functions. For example, on the way to the exhibition hall, visitors can pass by the museum bookstore, while the bookstore can provide simple catering as well.

右页图
罗马 21 世纪美术馆流线示意，依据扎哈·哈迪德的图解：主要脉络清晰的同时，保留了足够的并行线和支线。

Right
Diagram of circulation route at MAXXI Museum, Rome, based on works of Zaha Hadid. A clear circulation route goes with proper parallel lines and branches.

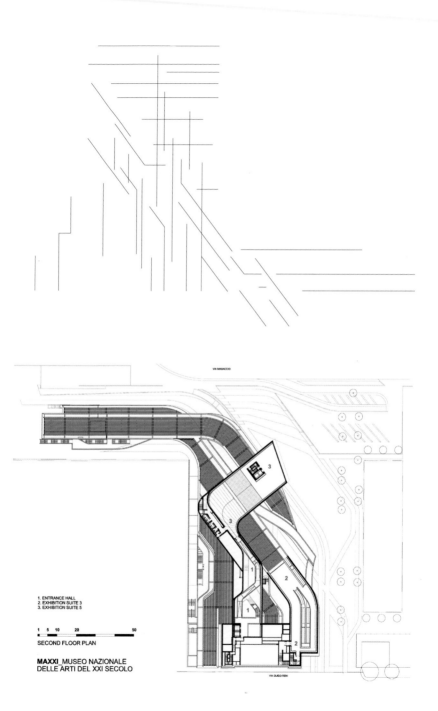

1. ENTRANCE HALL
2. EXHIBITION SUITE 3
3. EXHIBITION SUITE 5

1 5 10   20   50

SECOND FLOOR PLAN

MAXXI_MUSEO NAZIONALE
DELLE ARTI DEL XXI SECOLO

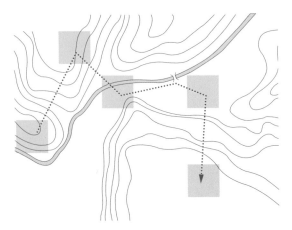

按景观形态导向

假设美术馆是一座花园，去往不同展厅便是去往不同的景观，看风景认路。

Landscape-driven Guide

Suppose that the art museum is a garden. Going to different exhibitions are like going to different scenery spots. Visitors find their way through the landscape.

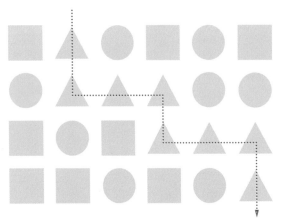

按空间品质导向

按照空间品质不同（大小方圆），即使没有任何标识也能轻松识途。

Spatial Character Guide

Different spatial characters (curvy, straight, linear, square, circular, etc.) offer easy ways for viewers to find their way even without road signs.

按空间命名导向

类似我们在餐厅去往不同的包房，空间性质相似只是名字不同，识别率不高。

Spatial Naming Guide

This is similar to guiding to different private rooms in a restaurant. The spaces have similar properties with the only difference being the names, resulting in low identifiability.

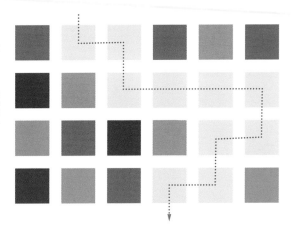

按城市结构逻辑导向

假设美术馆类似于一座城市，去往不同展厅便是去往不同的街区。

City Structural Logic Guide

Suppose the art museum is a city. The directions to exhibitions are similar to the directions to streets.

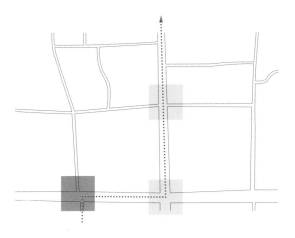

管理的议题
+
功能程序推导
Management
+
Program

○ 美术馆的一切功能围绕展览。
The museum program is centered around art exhibitions.

○ 现代城市生活的规模让展览管理日趋变得庞大复杂。
Museum management has become increasingly complicated in large modern cities.

○ 新颖的空间设计将为美术馆的管理带来方便。
Innovative spatial design facilitates museum management.

○ 建筑结构的模型取决于功能结构的模型。
The structure of architecture depends on the structure of functionality.

○ 公共活动、展览、服务像并行的轨道列车。
Public events, exhibitions and museum services go parallel to each other.

○ 工作流线的后面是功能项目的网络。
Streamlining of the museum program is network of functions.

○ "功能"有时需归并有时要推展，既简明又丰富。
"Functions" can be combined or extended, simple yet varied, to meet museum needs.

○ 功能最终转换成建筑空间的原则是：节约使用。
Functions ultimately shape architectural spaces on the basis of saving resources.

○ "功能"管理既需要严谨，也需要创新。
Management models for museum "functions" need to be both rational and innovative.

围绕展览
Exhibition-Centered

展览是美术馆的核心功能，研究美术馆的功能结构首先要从整体上研究美术馆中不同展览的关系，其次研究展览功能和其他建筑功能的关系。最直接的研究指标来自于美术馆一年中的不同展览展期的分布规律，不同内容的展览的具体要求，以及为实现这些展览美术馆各部门之间的配合情况。将抽象的指标和程序转换为整体空间关系也就是设计美术馆的过程。

细化的美术馆功能意味着明确的美术馆展览方向，它对设计的起点至关重要。

Holding exhibitions is the key function of the museum. Research on museum programs, first of all, ought to study the relationship between different exhibitions as a whole. Next, we need to understand the relationship between the exhibition functions and spatial functions within the museum. For this purpose, the most direct research targets include the regular annual calendars of museum exhibitions, which reveal the timing distribution of events, detailed requirements for different types of exhibitions, and the interactions among the supporting departments of the museum. Turning the abstract numbers and programs into a comprehensive spatial structure is the process of designing a museum.

A detailed museum function program means clarified museum exhibition goals – this provides a vital starting point for museum design.

1   单一作品主宰整个展厅
     A single work dominating the whole gallery

2   海量的作品同时位于一个博览会式美术馆空间内
     A large quantity of artworks displayed in the
     same museum space, like a trade fair

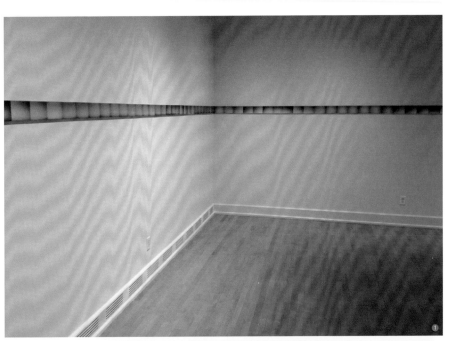

中国美术馆 2009 年各类展览分布图
The exhibitions of the National Art Museum of China, 2009

从本表中可以看到展览的次数，长度与频率在一年中的分布是不均匀的。由于不同展览门类对于展厅的要求不一样，现实中的美术馆使用对展览空间的灵活性提出了较高的要求。从图表中可以看到展览的不均匀分布，图中的空白区域和密集区域意味着一年中闲置的展厅和供不应求的展厅。这种状况往往是因为部分展厅无法适应于特定的展览要求而造成的。

In the diagram we see an uneven distribution of exhibitions in a year in terms of their number of times, duration and frequency. Due to different demands for display space by exhibitions of different types, in practical terms, art museums require highly flexible spaces. In the diagram, dense and empty areas indicate overused and unused galleries in different times of year. Low occupancy of galleries is often the result of certain museum spaces incapable of adapting to required uses.

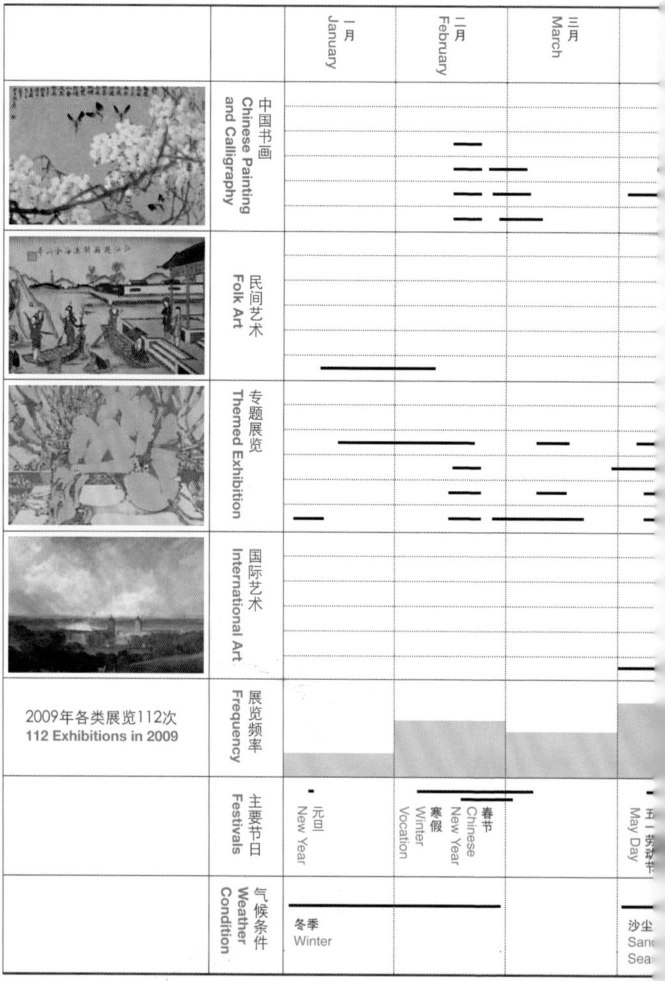

		一月 January	二月 February	三月 March
	中国书画 Chinese Painting and Calligraphy			
	民间艺术 Folk Art			
	专题展览 Themed Exhibition			
	国际艺术 International Art			
2009年各类展览112次 112 Exhibitions in 2009	展览频率 Frequency			
	主要节日 Festivals	元旦 New Year	春节 Chinese New Year / 寒假 Winter Vocation	五一劳动节 May Day
	气候条件 Weather Condition	冬季 Winter		沙尘 Sand Sea

220

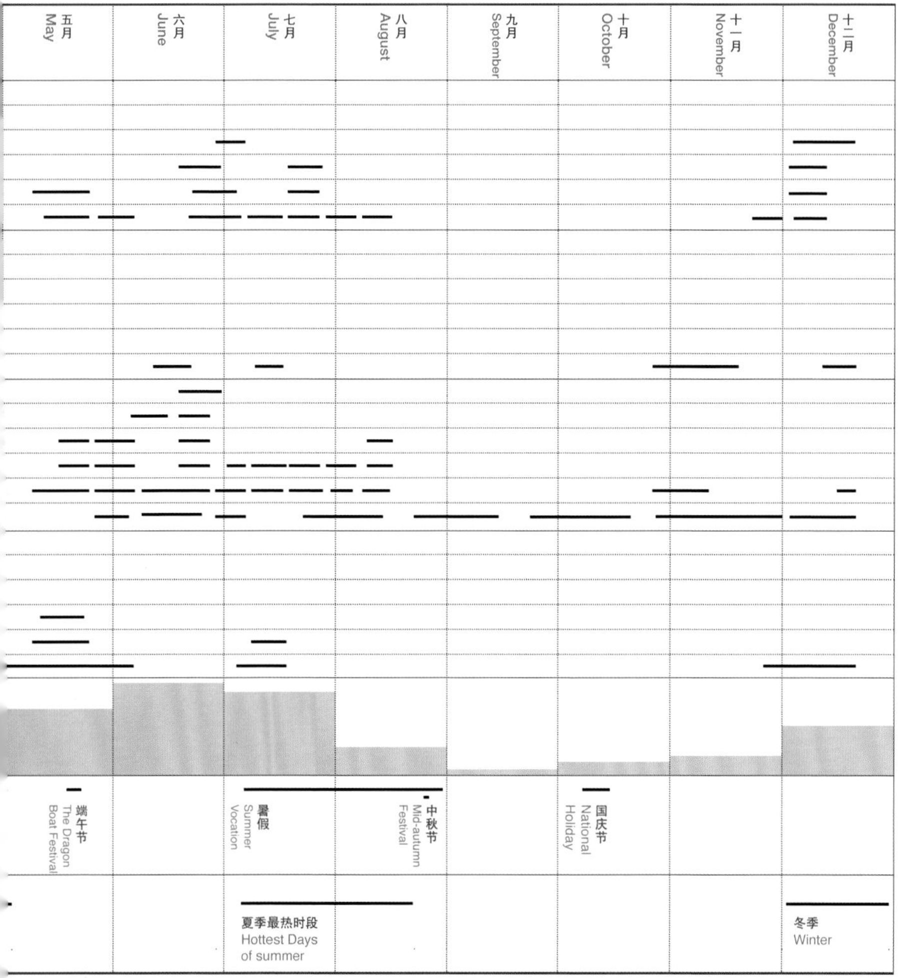

五月
May

六月
June

七月
July

八月
August

九月
September

十月
October

十一月
November

十二月
December

端午节
The Dragon
Boat Festival

暑假
Summer
Vacation

中秋节
Mid-autumn
Festival

国庆节
National
Holiday

夏季最热时段
Hottest Days
of summer

冬季
Winter

美术馆管理的空前压力来自巨型都市的观众流量

Museums Burdened with Flows of Metropolitan Visitors

中国城市普遍进入"超载"时期,例如北京是有着2114.8万人(截至2014年)常住人口的巨型都市,同时每年有大量外地访问者。从2011年开始中国国有美术馆已经全面向观众免费开放,中国美术馆8300平方米面积的展馆年均接待100万人,巨大的观众流量给美术馆未来的管理带来空前压力。

Most Chinese cities are entering an "age of overloading". For instance, Beijing is a megacity with a permanent population of 21.148 million (as of 2014). Moreover, it receives a colossal number of visitors from all over China and abroad every year. Since 2011 when the free admission policy was implemented in state-owned museums, the National Art Museum of China, with 8,300 square meters of exhibition spaces, has received an annual audience of 1 million. The large visitor flows have put unprecedented pressures on museum management.

1　2010年上海世界博览会:轻轨交通,步行的参观人群和核心展区之间的分流成为展场设计中最重要的管理问题之一。

　　Shanghai Expo 2010. Stratification of lifted rail transportation, pedestrian movement and fenced exhibition areas became one of the most important management issues for exhibition design.

2　2010年上海世界博览会:过长的排队等候时间中观众没有其他事可做。

　　Shanghai Expo 2010. The audience could do nothing but wait in queue.

3　2010年上海世界博览会英国馆:展场内外悬殊的观众密度。

　　Shanghai Expo 2010, UK pavilion. The density of the audience shows a dramatic contrast inside and outside of the pavilion.

空间设计 VS 管理模式
Space Design vs. Management Model

美术馆内部的人员密度分布同时受到设计意图和管理策略的影响，如同单向河道的水量和河道宽窄变化的关系，单一目的、单一方向的参观路径势必使得特定区域的美术馆空间的管理难度增大、停等时间延长、周边秩序紊乱；相反，如果能够设法使观众的流向均匀分布，分散观众活动的目的和方向，调动观众自我组织的能力，则可以减少停等时间，缓和周边压力。

The visitor density distribution inside the museum will be considerably influenced by the design and management strategies it adopts. Similar to the relationship between water volume and channel width for a one-way river, a one-way circulation route to a single exhibition destination will inevitably increase the difficulty in spatial management, prolonging waiting time and causing chaos in the surrounding areas. On the contrary, if the museum can provide more activity choices and direct the flow of visitors to various routes more evenly, reduction of visitors' waiting time and alleviation of crowdedness of surrounding areas will be achieved as the audience become self-organized.

边缘低密度区
Periphery Density: low

核心中密度区
Core Density: medium

核心高密度区
Core Density: high

主要人流方向
Major audience flow

主要展陈方向
Major exhibition orientation

G
展览区域
Exhibition area

 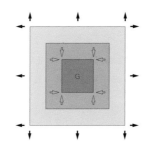 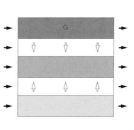

单向式

人流方向和观展方向完全一致，单向。

向心式

人流方向和观展方向一致，呈向心／离心分布。

过境式

人流方向和观展方向垂直。

Single Direction

The audience flow is consistent with the exhibition orientation. Single direction.

Center-oriented

The audience flow is consistent with the exhibition orientation. Concentric or centrifugal distribution.

Crossover

The audience flow is perpendicular to the exhibition orientation.

公共空间和展览空间的对接方式
The Relationship between Public and Exhibition Space

除了展览空间彼此的关系，公共空间和展览空间彼此的对接是美术馆
中最重要的结构关系。为了方便管理，美术馆的公共空间需要和展览
空间有合理的，多样化的对接方式。

- 二者的分工相对明确，接口清楚，互不干扰。
- 公共空间和展览空间可以互相渗透，彼此补充，提高整体的使用效率。
- 在满足总体容量的同时，公共空间的尺度应宜人、适度。

In addition to the relationship between gallery spaces, the interface among public space and exhibition space is pivotal to museums. For management convenience, public space and exhibition space need to be connected to each other in a reasonable and diversified way.

- Public space and exhibition space should have clear definitions regarding their exclusive functions, and would not interfere with each other. Their major interfaces should be unmistakably set.
- Public space and exhibition space should complement each other with some of their uses overlapping for a higher utility.
- Besides satisfying the museum's overall volume, public spaces in a museum should take a human scale and offer hospitable atmosphere.

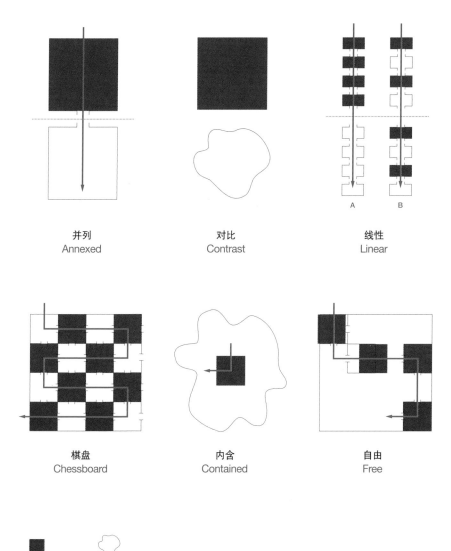

并列
Annexed

对比
Contrast

线性
Linear

A B

棋盘
Chessboard

内含
Contained

自由
Free

展览空间
Exhibition Space

公共空间
Public Space

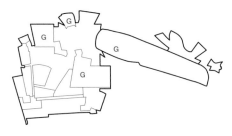

毕尔巴鄂古根海姆美术馆首层平面简图
First floor of Guggenheim Museum Bilbao

当代美术馆的大趋势是公共空间和展览空间不再有截然的界限，相应的，展览在何处开始和结束也没有以前那么明晰。

The general trend of contemporary museums sees no more definite boundaries between public space and exhibition space. As a result, it becomes vague regarding where an exhibition starts or ends.

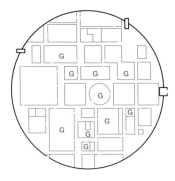

21 世纪金泽美术馆首层平面简图
First floor of 21 Century Museum of Contemporary Art, Kanazawa

北京中间美术馆展厅入口
Entrance to the gallery,
Inside-Out Museum, Beijing

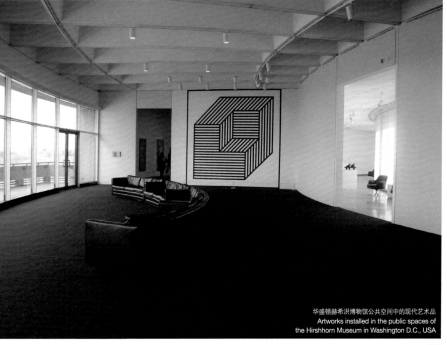

华盛顿赫希洪博物馆公共空间中的现代艺术品
Artworks installed in the public spaces of
the Hirshhorn Museum in Washington D.C., USA

功能结构的编制
Structuralizing a Museum Program

在中国，美术馆的机构设置与人员管理有着许多与西方同等美术馆不同的地方，只有充分了解这些差异性才能准确设置后台管理程序，并将它们转化为恰当的空间关系。因为尺度巨大和安全级别不一，如何设置新颖合理的空间管理程序，不至于体积过于庞大臃肿，并避免美术馆内部彼此间联络协同缺乏效率，是建筑师也需要考虑的问题。

在当代美术馆的活动中，展览已不再是唯一的项目，通过整合教育、娱乐和商业项目，美术馆成为国内外城市中重要的新型公共空间，担负很多额外功能。自 2011 年开始，包括中国美术馆在内的公立美术场馆向公众免费开放，美术馆观众的组成和数量发生了空前的变化。

In terms of institutional set-up and personnel management, Chinese museums are different from their western counterparts in many ways. A full understanding of these differences is essential to set up a proper backend management program and achieve appropriate spatial relationship. In a large building with varied security levels, it is the architect's responsibility to consider an innovative and reasonable space management program that creates synergies and efficiencies between different spaces, a program not too massive or ponderous to handle.

Exhibiting is no longer the sole function of the contemporary museum. By integrating its pedagogical, recreational and commercial program, museums have become a brand-new type of public space in cities worldwide. Starting in 2011, public Chinese museums, including NAMOC, have implemented free admission to the general public. The composition and scale of museum audience has been experiencing unprecedented changes.

中国美术馆网站结构图

网站虽然不是实在的建筑空间，但为美术馆的实际功能组成提供了结构化的图解，是一种虚拟的博物馆结构。

Structure, NAMOC's Official Website

Virtual web space is not equal to real space. Nonetheless, the website provides a diagram of the museum's function program, a map of virtual functional structure.

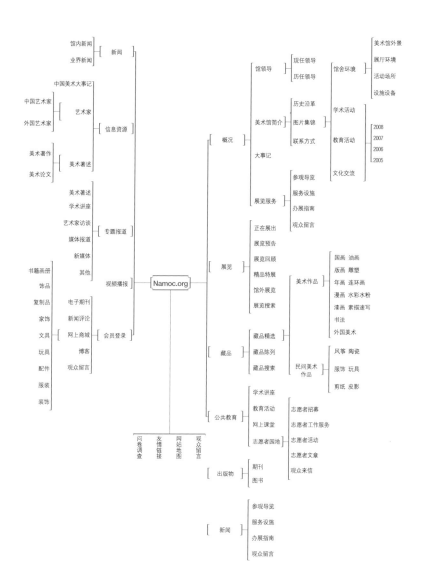

美术馆功能结构的地图
Functional Map of Museum Program

功能结构的地图既是美术馆主要功能的结构图,也可能是美术馆使用的"实战演习路线图"。传统的美术馆以展览为主线,前台的公共活动和后台的服务支持都是这条主线的从属,功能结构的地图并不是美术馆实际的空间结构,但是它基本反映空间组成的相互关系大局。管理流线中的各点在空间性质上相似,在空间关系上相邻。顺沿管理流线,各点共享空间资源,并可以最终回到美术馆的主要项目——展览。

A map of a museum's functional structure represents the major functions and activities of a museum, and simulates their actual uses. Exhibitions are the backbone of conventional art museums. Both the institution's public activities and its backstage services are ancillary functions connected to this backbone. The map of the museum's functional structure is not equal to its actual spatial structure; nevertheless, it configures the relationship between most spaces in the museum. Spots along the Management Flow line share similar spatial features and adjacent spatial relationship. Along the Management Flow line, they share spatial resources and eventually contribute to the core functionality of the museum – the exhibition.

右页图

传统的美术馆功能结构往往是树状的,均匀分布,在实际使用中则有必要区分其层次并使它们按先后顺序串联起来。

Right

The functional structure of a traditional museum is commonly a tree-type diagram with evenly distributed branches. In practice, it is necessary to clarify the levels and connect the functions in order and in sequence.

中国美术馆员工结构图
Office Structure Diagram of NAMOC Staff Composition

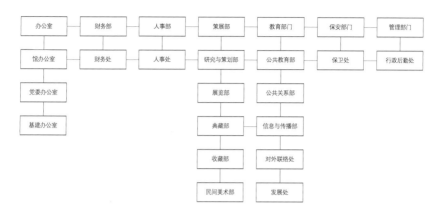

今日美术馆结构图
Structure Diagram of the Today Art Museum

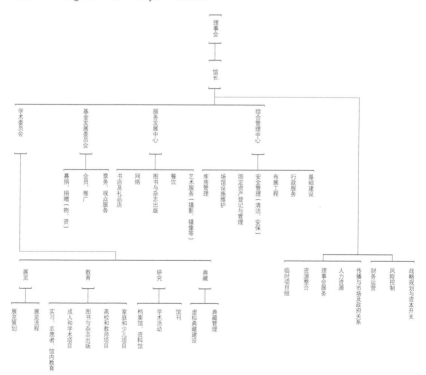

主要功能项目大类
Items of a Museum Program

基于上述美术馆功能结构的地图，我们可以综合汇编出 11 类美术馆建筑程序中的主要活动单元（"项目"，详见第 238 页）。它们按照从公共空间到私密空间的层次线性排列。这些基本"项目"的清单只是反映空间的组成情况，并不代表实际的空间关系。在这个层次上，建筑师需要考虑的是每一项目的功能性质（展览、服务等），与其他功能的空间关系（相邻、相隔等），项目的实际空间感受（大小、强度等）。

Based on the above-mentioned map of the museum's functional structure, we have compiled 11 categories of major types of activities ("items", see p.238) for museum architectural programming. They are ordered linearly from the more public to the more private in terms of space. These essential "items" reflect spatial composition, but do not represent actual spatial relationships. Architects need to take into account functions of each item (such as exhibition, service, etc.), its spatial relationship with other functions (whether it is near or far) and the actual spatial perception (such as size, intensity, etc.).

1. 咖啡厅
    Cafe
2. 接待
    Reception
3. 商店
    Shop
4. 图书馆
    Library
5. 设计室
    Studio
6. 剧院
    Theatre
7. 实验室
    Laboratory
8. 贮存空间
    Storage

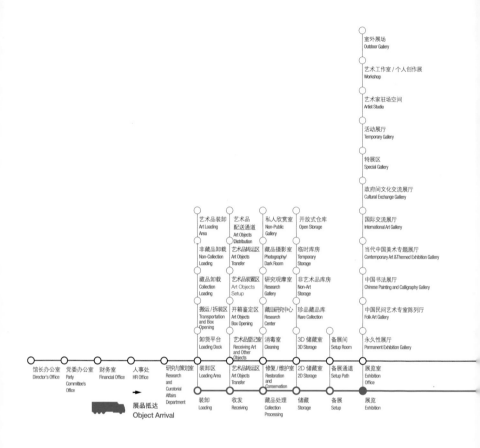

室外展场
Outdoor Gallery

艺术工作室 / 个人创作展
Workshop

艺术家驻场空间
Artist Studio

活动展厅
Temporary Gallery

特展区
Special Gallery

政府间文化交流展厅
Cultural Exchange Gallery

国际交流展厅
International Art Gallery

当代中国美术专题展厅
Contemporary Art & Themed Exhibition Gallery

中国书法展厅
Chinese Painting and Calligraphy Gallery

中国民间艺术专室陈列厅
Folk Art Gallery

永久性展厅
Permanent Exhibition Gallery

艺术品装卸 | 艺术品 | 私人欣赏室 | 开放式仓库
Art Loading Area | 配送通道 Art Objects Distribution | Non-Public Gallery | Open Storage

非藏品卸载 | 艺术品转运区 | 藏品摄影室 | 临时库房
Non-Collection Loading | Art Objects Transfer | Photography/ Dark Room | Temporary Storage

藏品卸载 | 艺术品装置区 | 研究观摩室 | 非艺术品库房
Collection Loading | Art Objects Setup | Research Gallery | Non-Art Storage

搬运 / 拆装区 | 开箱鉴定区 | 藏品研究中心 | 珍品藏品库
Transportation and Box Opening | Art Objects Box Opening | Research Center | Rare Collection

卸货平台 | 艺术品登记区 | 消毒室 | 3D 储藏室 | 备品间
Loading Dock | Receiving Art and Other Objects | Cleaning | 3D Storage | Setup Room

馆长办公室 | 党委办公室 | 财务室 | 人事处 | 研究与策划馆 | 装卸区 | 艺术品转运区 | 修复 / 维护室 | 2D 储藏室 | 备展通道 | 展览室
Director's Office | Party Committee's Office | Financial Office | HR Office | Research and Curatorial Affairs Department | Loading Area | Art Objects Transfer | Restoration and Conservation | 2D Storage | Setup Path | Exhibition Office

展品抵达
Object Arrival

装卸 | 收发 | 藏品处理 | 储藏 | 备展 | 展览
Loading | Receiving | Collection Processing | Storage | Setup | Exhibition

开架图书馆 | 闭架图书馆 | 善本图书馆 | 电子阅览室
Library Stack | Library Circulation | Rare Collection | E-Reading Room

电脑教室
Computer Classroom

就像任何地图一样，美术馆功能结构的地图既提供了若干明晰的"主要线路"，又描述了若干副轴和旁枝。

Just like any map, the map of the museum program provides both "thoroughfares" and branches.

展览项目
Exhibition

展品抵达
Art Object Flow

管理轴线
Administration Spine

后台服务
Back-end Service

观众抵达
Audience Flow

展览轴线
Exhibition Spine

公共活动
Public Events

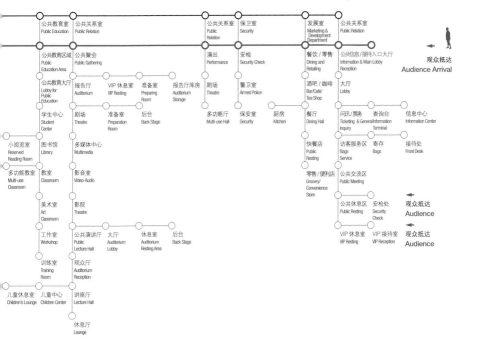

公共教育室 Public Education　公共关系室 Public Relation　公共关系室 Public Relation　保卫室 Security　发展室 Marketing & Development Department　公共关系室 Public Relation

公共教育区域 Public Education Area　公共聚会 Public Gathering　演出 Performance　安检 Security Check　餐饮/零售 Dining and Retailing　公共信息/接待入口大厅 Information & Main Lobby Reception　观众抵达 Audience Arrival

公共教育大厅 Lobby for Public Education　报告厅 Auditorium　VIP 休息室 VIP Resting　准备室 Preparing Room　报告厅库房 Auditorium Storage　剧场 Theatre　警卫室 Armed Police　酒吧/咖啡 Bar/Café/Tea Shop　大厅 Lobby

学生中心 Student Center　剧场 Theatre　准备室 Preparation Room　后台 Back Stage　多功能厅 Multi-use Hall　保安室 Security　厨房 Kitchen　餐厅 Dining Hall　问讯/票务 Ticketing & General Inquiry　查询台 Information Terminal　信息中心 Information Center

小阅览室 Reserved Reading Room　图书馆 Library　多媒体中心 Multimedia　快餐店 Public Resting　访客服务区 Bags Service　寄存 Bags　接待处 Front Desk

多功能教室 Multi-use Classroom　教室 Classroom　影音室 Video-Audio　零售/便利店 Grocery/Convenience Store　公共交流区 Public Meeting

美术室 Art Classroom　影院 Theatre　公共休息区 Public Resting　安检处 Security Check　观众抵达 Audience

工作室 Workshop　公共演讲厅 Public Lecture Hall　大厅 Auditorium Lobby　休息室 Auditorium Resting Area　后台 Back Stage　VIP 休息室 VIP Resting　VIP 接待室 VIP Reception　观众抵达 Audience

训练室 Training Room　观众厅 Auditorium Reception

儿童休息室 Children's Lounge　儿童中心 Children Center　讲座厅 Lecture Hall

休息厅 Lounge

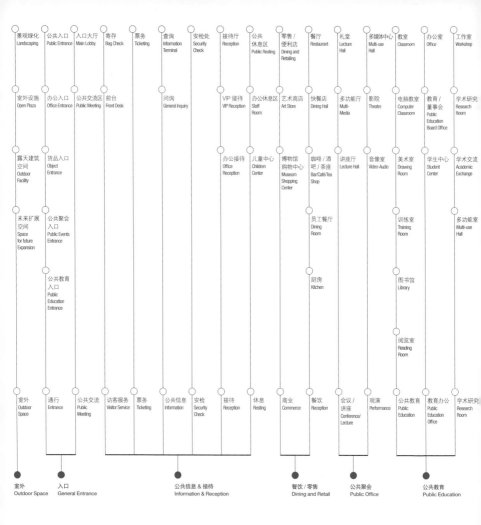

景观绿化 Landscaping	公共入口 Public Entrance	入口大厅 Main Lobby	寄存 Bag Check	票务 Ticketing	查询 Information Terminal	安检处 Security Check	接待厅 Reception	公共休息区 Public Resting	零售 / 便利店 Retail and Retailing	餐厅 Restaurant	礼堂 Lecture Hall	多媒体中心 Multi-use Hall	教室 Classroom	办公室 Office	工作室 Workshop
室外设施 Open Plaza	办公入口 Office Entrance	公共交流区 Public Meeting	前台 Front Desk		问询 General Inquiry		VIP 接待 VIP Reception	办公休息区 Staff Room	艺术商店 Art Store	快餐店 Dining Hall	多功能厅 Multi-Media	影院 Theatre	电脑教室 Computer Classroom	教育 / 董事会 Public Education Board Office	学术研究 Research Room
露天建筑空间 Outdoor Facility	货品入口 Object Entrance						办公接待 Office Reception	儿童中心 Children Center	博物馆购物中心 Museum Shopping Center	咖啡 / 酒吧 / 茶座 Bar/Café/Tea Shop	讲座厅 Lecture Hall	音像室 Video-Audio	美术室 Drawing Room	学生中心 Student Center	学术交流 Academic Exchange
未来扩展空间 Space for future Expansion	公共聚会入口 Public Events Entrance									员工餐厅 Dining Room			训练室 Training Room		多功能室 Multi-use Hall
	公共教育入口 Public Education Entrance									厨房 Kitchen			图书馆 Library		
													阅览室 Reading Room		
室外 Outdoor Space	通行 Entrance	公共交流 Public Meeting	访客服务 Visitor Service	票务 Ticketing	公共信息 Information	安检 Security Check	接待 Reception	休息 Resting	商业 Commerce	餐饮 Reception	会议 / 讲座 Conference/ Lecture	观演 Performance	公共教育 Public Education	教育办公 Public Education Office	学术研究 Research Room

● 室外 Outdoor Space	● 入口 General Entrance	● 公共信息 & 接待 Information & Reception	● 餐饮 / 零售 Dining and Retail	● 公共聚会 Public Office	● 公共教育 Public Education

本表列出了供建筑师参考的美术馆中的 11 类主要活动单元，这些活动单元都是建筑中不可拆分、无法互相替代的成分，它体现了一座现代的中国美术馆所提供的基本功能。按从公共到私密的分类而列出的这些功能单元只是反映空间的大致组成情况，并不代表实际的空间关系。

This diagram lists, for architects, 11 major types of activities held in the museum, which are irreducible and irreplaceable components. They specify the basic functions of a modern art museum. These functional units, ordered from the more public to the more private, are categorized to reflect the functional composition of spaces, without necessarily reflecting the actual spatial relationship.

区 藏品室 Storage | 各部门办公室 Office | 研究与策划 Research and Curatorial Affairs Department | 会议室 Conference Room | 值班室 Security Post | 更衣间/淋浴间 Change Room/ Shower Room | 修复室 Mounting | 藏品摄影 Photography /DarkRoom | 实验室 Laboratory | 鉴定室 Inspection Room | 装卸区 Loading Area | 动力房 Power Room

通道 开放仓库 Open Storage | 档案室 Archive Department | 信息传播 Information Department | 会客室 Guest Room | 休息室 Restroom | 服务间/清洁区 Cleaning Service | 维护室 Conservation | 资料室 Archive | 研究室 Research Room | 开箱鉴定 Art Objects Box Opening | 机械室 Mechanic Room

办公储藏 Office Storage | 后勤 Logistics Department | 对外联络 International Department | 视听室 Video/Audio Room | 武警营房 Armed Police Dormitory | 医务室 Clinic | 修裱室 Restoration | 图像服务 Image Service | 工作室 Workshop | 入库登记 Objects Register | 设备间 Equipment Room

教育储藏 Education Storage | 消毒室 Cleaning | 接收包装 Art Objects Packaging | 洗衣房 Laundry

临时存储 Temporary Storage for Art Object | 配送转运 Art Objects Transfer

工具储藏 Tools

贮藏 Storage | 办公 Administration | 作业 Professional | 会议 Conference | 安保 Security Department | 卫生清洁 Cleaning Service | 修复维护 Restoration and Conservation | 媒体资料 Media | 实验 Laboratory | 藏品研究 Research | 收发 Receiving | 技术设施 Facilities/ Structures

贮藏 Storage | 办公 Office | 藏品处理 Collection Processing | 收发 Receiving | 技术设施 Facilities/Structures

239

将个别的功能项目整合成系统的空间关系
Integrating Individual Functions into a Spatial System

原则上，建筑程序中初次列出的每一项目均是不可拆分、不相重复的。实际上在明确建筑基本功能大类区分的前提下，建议每大类的项目适当重叠，互相衔接，使得建筑使用的空间分布混合和多样化。多样化的功能整合是空间设计的关键。这一过程也是建筑功能向着建筑空间转化的开始，对于项目间重叠衔接关系的不同理解，反映了不同设计师设计的空间模型差异。

在管理层面上，每一功能项目可能是孤立的，也有可能彼此重复，编制合理的建筑程序，需要把管理层面的功能要求整合成系统的空间关系。创造性的空间设计需要灵活地处理不同项目的关系：首先，同一空间的属性可以发生变化满足不同的要求；其次，不同空间之间可以发生这样那样的功能重叠，以减少重复建设的部分，提高空间效率。

In principle, all items listed in architecture program are irreducible or irreplaceable. In reality, if the major functional categories are explicitly defined, certain items in each category can be overlapped to some degree, to make space utilization diversified and multifunctional. The integration of diversified functions is critical in spatial design. This integration process initiates the transformation from architectural function into architectural space; different understandings of the overlapping relationship between items reflect the difference between spatial models by different architects.

At the management level, each functional requirement could be isolated or overlapping. Reasonable architectural programming aims to integrate separate functions into a systematic spatial organization. Innovative spatial design should deal with the functional relationship flexibly: the same space can be converted to accommodate multiple functions, and different spaces can share the same function so as to reduce redundancy and increase the efficient use of space.

同一空间中不同性质空间的
转换和区分

Conversion of Functional
Use in the Same Space

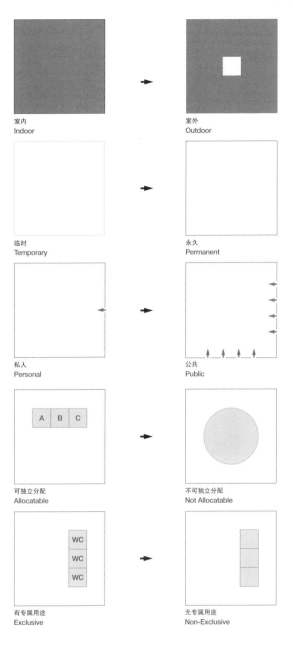

室内
Indoor

室外
Outdoor

临时
Temporary

永久
Permanent

私人
Personal

公共
Public

可独立分配
Allocatable

不可独立分配
Not Allocatable

有专属用途
Exclusive

无专属用途
Non-Exclusive

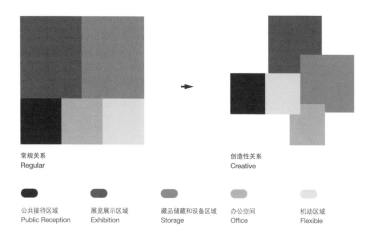

常规关系
Regular

创造性关系
Creative

公共接待区域
Public Reception

展览展示区域
Exhibition

藏品储藏和设备区域
Storage

办公空间
Office

机动区域
Flexible

不同功能的重叠和同一空间的转化

"重叠"指的是同一功能可以在不同空间单元中出现,呈现不同的空间性质。例如,处于入口处的休息厅可以允许一定的环境噪音,而处于画廊之间的某些休息厅则需要安静。

"转化"指的是同一空间单元可以用作别的用途,例如,阅览室和剧院属于独立分配的、永久使用的公共空间,但是它们很难改为其他用途;相反,工作室和小会议厅是多功能空间,可以用来工作、自习,也可以用来举办性质更活跃的公共活动。

Overlapping and Convertible Functions/Spaces

"Overlapping" refers to the same type of function used in different spaces while showing dissimilar spatial properties. For example, if the resting area is located at the entrance, ambient noise can be allowed to some degree, while the resting area between the galleries should require silence.

"Convertible" refers to the possibility of a space being changed to serve different functions. For instance, the reading rooms and the theatre are designated with exclusive functions permanently, as it is hard to transform either of them for other uses. In contrast, the studio and the meeting room are multi-functional spaces, and can be used to work, study, or even to host some active public events.

当代美术馆力图使得纯"功能"性的空间单元彼此混合，使得每一大类功能在浓缩加强核心样式的同时，其空间范围又往往有所扩充。例如，新古典主义平面中静态的，区分前后、面积有限的单一"入口"已经被大类的"入口区域"取代，接待台被整合为一件多功能家具或是退到空间一侧。

Contemporary art museums tend to mix functional components. Certain cored functions are enhanced while their spatial realms are expanded and reconstructed as new types. For instance, in the Neo-Classical plan, the single small entrance that pinpoints the major building outlet has been replaced with the larger "welcoming area". The classical reception desk has been redesigned as a multi-functional furniture or set aside in space.

巴塞尔的舒拉格美术馆入口
The entrance of Schaulager in Basel

舒拉格美术馆接待区
The reception area of Schaulager

日本美秀美术馆展厅入口处公共区域
The public area at the entrance of Miho Museum

德国埃森的弗柯望美术馆前台区域
The reception area at Folkwang Museum, Essen, Germany.

将建筑的功能项目落实为建筑的实际组成

Actualizing the Museum Program through Architectural Components

将功能项目的不同关系呈现在一个三维的图解里，就显示出在建筑功能项目落实为建筑实际组成时出现的实际情况。图中可以看出在将抽象的建筑功能转换为具体的空间结构时，不同的建筑师会有多么不同的策略。

大致说来，在水平和垂直两个方向上，建筑的空间结构都应该遵循博物馆建筑的公共空间 – 展览空间 – 办公空间三段式功能原则。换而言之，由公共的开敞的城市空间逐渐过渡到较为肃穆和相对封闭的画廊空间，再由庄重的展示环境过渡到积极的但边界封闭的工作空间。

美术馆的空间组成粗略地遵循着三段式原则。同时，一些在传统美术馆中不曾强调的功能在当代空间表现上具有强调和延展的可能，它们或许打破了上述的三段式功能原则，但仍然遵从上述功能空间的重叠和转化的原则。

Three-dimensional diagrams exemplify some possible scenarios we may face in the process of actualizing functional units in space. These schemes illustrate different strategies adopted by different architects in their efforts to implement abstract museum programs into actual spatial layout.

Generally speaking, the spatial structure of the museum building should adhere to the tri-sectional function principle both horizontally or vertically: public space – exhibition space – work space. In other words, there should be a smooth transition from a public, open, urban space to a comparatively solemn and enclosed gallery space, and then from the solemn gallery space to the more dynamic yet self-enclosed working space.

Most museums follow the tri-sectional function principle to some degree. Contemporary museums, however, often feature spatial configurations not emphasized in traditional museums, enhancing and expanding certain categories in its program. They may diverge from the tri-sectional principle, but remain consistent with the above-mentioned overlapping and convertible principles.

相同功能在不同平面上的连接

把空间结构中性质相同的部分彼此连接，往往就自然构成了建筑物内部的核心流线。

Same function on different building levels

Consistent design for spaces with the same functions across different building levels makes naturally the museum's core circulation routes.

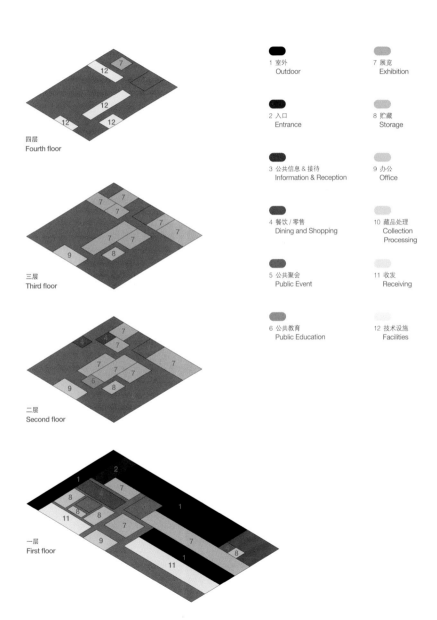

四层
Fourth floor

三层
Third floor

二层
Second floor

一层
First floor

1 室外
Outdoor

2 入口
Entrance

3 公共信息 & 接待
Information & Reception

4 餐饮 / 零售
Dining and Shopping

5 公共聚会
Public Event

6 公共教育
Public Education

7 展览
Exhibition

8 贮藏
Storage

9 办公
Office

10 藏品处理
Collection Processing

11 收发
Receiving

12 技术设施
Facilities

编制建筑程序
Creating an Architectural Program

编制建筑程序（亦称建筑策划）意味着建筑设计的正式开始。通过预设管理使用的模式，并且将建造的建筑单元细化和现实化，建筑程序汇编了博物馆的主要功能。更重要的是，一个具体的建筑程序定义了不同尺寸的空间是如何设置和组合在一起的。遵从这样的建筑程序，并且以适当的空间样式将建筑程序转为现实，设计师就可以设计出一个既有创造性，也有理性和现实基础的博物馆。

一个建筑程序包含两类要素：以抽象形式呈现的功能结构，以及将这些功能结构以建筑学方式组织在一起的可分析"类型"；两者都可以帮助更好地理解艺术展览的功能和博物馆的城市文脉。

Compiling an architectural program is the formal starting point of architectural design. An architectural program specifies major museum functions by presetting its model of management and uses and by detailing the type and quantity of architectural components to build. Most importantly, a specific architectural program defines how spaces of different sizes are configured and combined. Following such a program and its actualization in appropriate spatial forms, architects are able to design a museum that is both creative and sound with a rational and practical basis.

An architectural program is made up of two elements: a functional structure (in abstract forms) in combination with analytical "types" (in architectural principle) that help to organize the functional structures. Both can help to illuminate a deep understanding of the art exhibition function and the urban context of the museum.

右图
不同的美术馆设计、美术馆定位、建筑功能和展览策略
的组合带来不同的平面的设计。

Right

Different combinations of museum design, museum positioning, architectural function and exhibition strategy result in different museum layout of the plan.

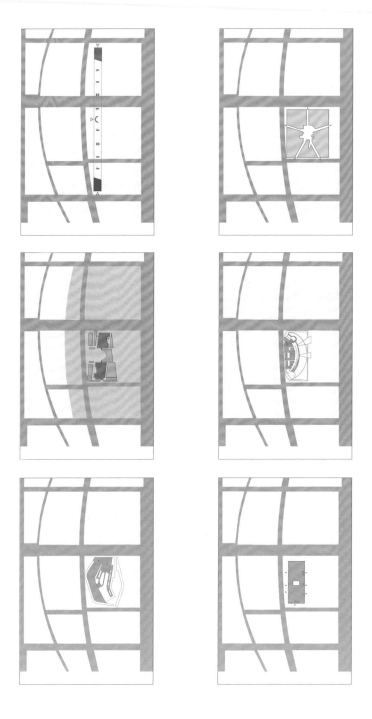

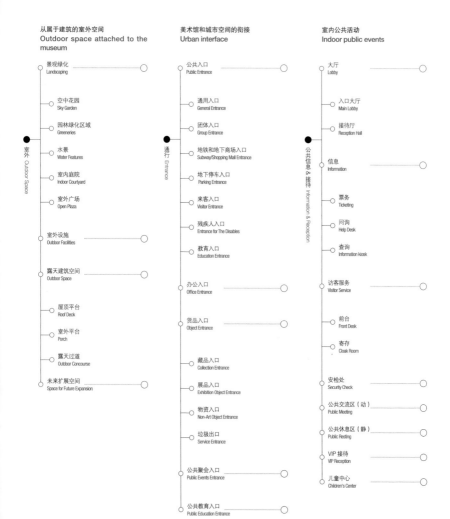

从属于建筑的室外空间
Outdoor space attached to the museum

景观绿化
Landscaping

空中花园
Sky Garden

园林绿化区域
Greeneries

水景
Water Features

室内庭院
Indoor Courtyard

室外广场
Open Plaza

室外设施
Outdoor Facilities

露天建筑空间
Outdoor Space

屋顶平台
Roof Deck

室外平台
Porch

露天过道
Outdoor Concourse

未来扩展空间
Space for Future Expansion

室外 Outdoor Space

美术馆和城市空间的衔接
Urban interface

公共入口
Public Entrance

通用入口
General Entrance

团体入口
Group Entrance

地铁和地下商场入口
Subway/Shopping Mall Entrance

地下停车入口
Parking Entrance

来客入口
Visitor Entrance

残疾人入口
Entrance for The Disables

教育入口
Education Entrance

办公入口
Office Entrance

货品入口
Object Entrance

藏品入口
Collection Entrance

展品入口
Exhibition Object Entrance

物资入口
Non-Art Object Entrance

垃圾出口
Service Entrance

公共聚会入口
Public Events Entrance

公共教育入口
Public Education Entrance

通行 Entrance

室内公共活动
Indoor public events

大厅
Lobby

入口大厅
Main Lobby

接待厅
Reception Hall

信息
Information

票务
Ticketing

问询
Help Desk

查询
Information kiosk

访客服务
Visitor Service

前台
Front Desk

寄存
Cloak Room

安检处
Security Check

公共交流区（动）
Public Meeting

公共休息区（静）
Public Resting

VIP 接待
VIP Reception

儿童中心
Children's Center

公共信息 & 接待 Information & Reception

安全级别
Security Level

● 非公共有展品区
Non-public with artworks

○ 非公共无展品区
Non-public without artworks

● 公共有展品区
Public with artworks

○ 公共无展品区
Public without artworks

餐饮 / 零售
Catering and shopping

零售 / 便利店
Convenience Store

艺术商店
Art Store

纪念品商店
Souvenir/Book Store

艺术品商店
Art Objects Store

博物馆购物中心
Museum Shop

商店储藏
Store Storage

餐厅
Restaurant

中式餐厅
Chinese Restaurant

西式餐厅
Western-Style Restaurant

快餐店
Food Court

酒吧 / 咖啡 / 茶座
Bar/Café/Tea

厨房
Kitchen

餐饮 / 零售 Catering and Shopping

会议 / 讲座 / 演出
Conference/lecture/performance

多功能厅
Multi-use Hall

多媒体中心 / 音像室
Multimedia/Video-Audio

影院
Theatre

礼堂
Auditorium

大堂
Lobby

休息室
Auditorium Lobby

后台
Back-Stage

讲座厅
Lecture Hall

观众休息厅
Audience Reception

公共交流 Public Exchange

美术馆和城市空间的衔接
Public education

大厅
Lobby for Public Education

教育办公室
Public Education Office

图书阅览
Library

开架图书馆
Library with Open Stacks

闭架图书馆
Library (Circulation Only)

善本图书馆
Rare Collection

电子阅览室
E-Reading Room

小阅览室
Reserved Reading Room

教室
Classroom

教室
Classroom

电脑教室
Computer Classroom

多功能教室
Multi-use Classroom

美术室
Drawing Room

训练室
Training Room

工作室
Workshop

教育 / 董事会
Public Education Board Office

学术研究
Research Room

学术交流
Academic Exchange

学生中心
Student Center

公共教育 Public Education

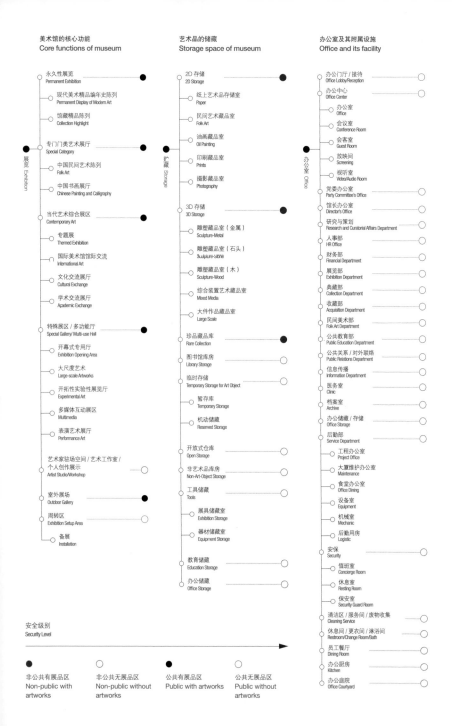

美术馆的核心功能 Core functions of museum	艺术品的储藏 Storage space of museum	办公室及其附属设施 Office and its facility

美术馆的核心功能 / Core functions of museum

展览 Exhibition

永久性展览 Permanent Exhibition
- 现代美术精品编年史陈列 Permanent Display of Modern Art
- 馆藏精品陈列 Collection Highlight

专门门类艺术展厅 Special Category
- 中国民间艺术陈列 Folk Art
- 中国书画展厅 Chinese Painting and Calligraphy

当代艺术综合展区 Contemporary Art
- 专题展 Themed Exhibition
- 国际美术馆馆际交流 International Art
- 文化交流展厅 Cultural Exchange
- 学术交流展厅 Academic Exchange

特殊展区 / 多功能厅 Special Gallery/ Multi-use Hall
- 开幕式专用厅 Exhibition Opening Area
- 大尺度艺术 Large-scale Artworks
- 开拓性实验性展览厅 Experimental Art
- 多媒体互动展区 Multimedia
- 表演艺术展厅 Performance Art

艺术家驻场空间 / 艺术工作室 / 个人创作展示 Artist Studio/Workshop

室外展场 Outdoor Gallery

周转区 Exhibition Setup Area
- 备展 Installation

艺术品的储藏 / Storage space of museum

储藏 Storage

2D 存储 2D Storage
- 纸上艺术品存储室 Paper
- 民间艺术藏品室 Folk Art
- 油画藏品室 Oil Painting
- 印刷藏品室 Prints
- 摄影藏品室 Photography

3D 存储 3D Storage
- 雕塑藏品室（金属）Sculpture-Metal
- 雕塑藏品室（石头）Sculpture-stone
- 雕塑藏品室（木）Sculpture-Wood
- 综合装置艺术藏品室 Mixed Media
- 大件作品藏品室 Large Scale

珍品藏品库 Rare Collection

图书馆库房 Library Storage

临时存储 Temporary Storage for Art Object
- 暂存库 Temporary Storage
- 机动储藏 Reserved Storage

开放式仓库 Open Storage

非艺术品库房 Non-Art-Object Storage

工具储藏 Tools
- 展具储藏室 Exhibition Storage
- 器材储藏室 Equipment Storage

教育储藏 Education Storage

办公储藏 Office Storage

办公室及其附属设施 / Office and its facility

办公室 Office

办公门厅 / 接待 Office Lobby/Reception

办公中心 Office Center
- 办公室 Office
- 会议室 Conference Room
- 会客室 Guest Room
- 放映间 Screening
- 视听室 Video/Audio Room

党委办公室 Party Committee's Office
馆长办公室 Director's Office
研究与策划 Research and Curatorial Affairs Department
人事部 HR Office
财务部 Financial Department
展览部 Exhibition Department
典藏部 Collection Department
收藏部 Acquisition Department
民间美术部 Folk Art Department
公共教育部 Public Education Department
公共关系 / 对外联络 Public Relations Department
信息传播 Information Department
医务室 Clinic
档案室 Archive
办公储藏 / 存储 Office Storage

后勤部 Service Department
- 工程办公室 Project Office
- 大厦维护办公室 Maintenance
- 食堂办公室 Office Dining
- 设备室 Equipment
- 机械室 Mechanic
- 后勤用房 Logistic

安保 Security
- 值班室 Concierge Room
- 休息室 Resting Room
- 保安室 Security Guard Room

清洁区 / 服务间 / 废物收集 Cleaning Service
休息间 / 更衣间 / 淋浴间 Restroom/Change Room/Bath
员工餐厅 Dining Room
办公厨房 Kitchen
办公庭院 Office Courtyard

安全级别 / Security Level

- ● 非公共有展品区 Non-public with artworks
- ○ 非公共无展品区 Non-public without artworks
- ● 公共有展品区 Public with artworks
- ○ 公共无展品区 Public without artworks

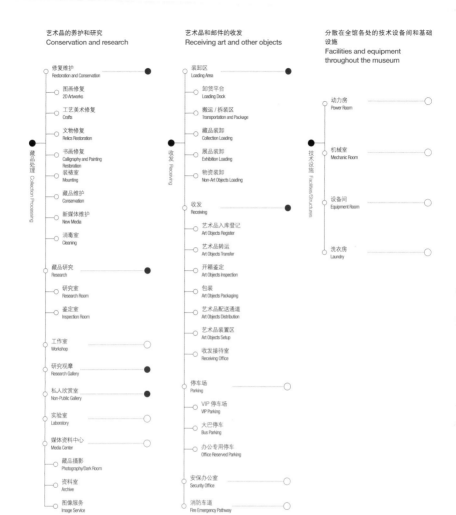

艺术品的养护和研究
Conservation and research

修复维护
Restoration and Conservation

图画修复
2D Artworks

工艺美术修复
Crafts

文物修复
Relics Restoration

书画修复
Calligraphy and Painting
Restoration

装裱室
Mounting

藏品维护
Conservation

新媒体维护
New Media

消毒室
Cleaning

藏品研究
Research

研究室
Research Room

鉴定室
Inspection Room

工作室
Workshop

研究观摩
Research Gallery

私人欣赏室
Non-Public Gallery

实验室
Labontory

媒体资料中心
Media Center

藏品摄影
Photography/Dark Room

资料室
Archive

图像服务
Image Service

藏品处理 Collection Processing

艺术品和邮件的收发
Receiving art and other objects

装卸区
Loading Area

卸货平台
Loading Dock

搬运 / 拆装区
Transportation and Package

藏品装卸
Collection Loading

展品装卸
Exhibition Loading

物资装卸
Non-Art Objects Loading

收发
Receiving

艺术品入库登记
Art Objects Register

艺术品转运
Art Objects Transfer

开箱鉴定
Art Objects Inspection

包装
Art Objects Packaging

艺术品配送通道
Art Objects Distribution

艺术品装置区
Art Objects Setup

收发接待室
Receiving Office

停车场
Parking

VIP 停车场
VIP Parking

大巴停车
Bus Parking

办公专用停车
Office Reserved Parking

安保办公室
Security Office

消防车道
Fire Emergency Pathway

收发 Receiving

分散在全馆各处的技术设备间和基础设施
Facilities and equipment throughout the museum

动力房
Power Room

机械室
Mechanic Room

设备间
Equipment Room

洗衣房
Laundry

技术设施 Facilities/Structures

创意的空间
Creative Museum Space

清晰的建筑程序排布并不妨碍产生丰富多变的美术馆空间。归根结底，"功能只有在它找到一种好的表达形式的时候才成其为功能"（彼得·艾森曼）。空间产生新创意的变数来自于打破常规的功能排布，带来对一般功能程序的重新审视。

A succinct architectural program does not necessarily lead to boring and repetitive museum spaces. As Peter Eisenman said, "a function becomes function only when it finds a good expression." Innovative museum spaces come from the unconventional arrangement of functions which allows for a reexamination of the regular museum programs.

1/2   李柏斯金的柏林犹太人博物馆，其线性平面象征着犹太人的流亡之旅。

The Jewish Museum in Berlin designed by Daniel Libeskind. The zigzagging museum plan symbolizes the journey of the Jews in exile.

3/4   贝聿铭设计的日本美秀美术馆，来访者经过长途跋涉才能到达美术馆，这种安排也是参观体验的一部分。

Miho Museum, Japan, designed by I. M. Pei. Visitors need to take a long way to arrive at the museum – that is a special arrangement to experience the museum.

创意的概念
长长的穿越之旅："行动"本身成为博物馆的主线索。

Creative in Museum Concept
In a long journey, "action" becomes a major feature of the
museum.

创意的质感
构成展览空间的不同透明度可以产生创新变化。

Creative in Museum Texture
Forming a gradient of transparency in museum spaces can lead to innovation.

中国人民大学博物馆二层展厅，不同透明度的引入。
Different transparencies introduced in the space, as seen in the
2nd-level gallery, Renmin University of China Museum, Beijing.

比利时布鲁塞尔 2009 年欧罗巴利亚中国艺术节 "活的中国园林"。展场设计里有通透性的室内分隔。
"Chinese Gardens for Living" exhibition at Europalia Art Festival 2009, Brussels, Belgium.
Translucent interior divisions in the gallery design.

创意的空间

不再是单纯的"盒子",空间自身就是一种内在的造型因素和独特的艺术品。

Creative in Museum Space

No longer a pure cube, the exhibition space itself is now a unique artwork of its own form-making logic.

中央美院美术馆自由形态的展览空间。
Free-form exhibition space at Central Academy of Fine Arts (CAFA) Art Museum.

BIG 设计的 2010 上海世界博览会丹麦馆,螺旋形展览空间。
Denmark Pavilion (designed by BIG), Shanghai Expo 2010. Spiral exhibition space.

第八章

可建造的美术馆
The Buildable Museum

○ 美术馆建筑的主要专业设备需求是照明和展示。

Major museum facilities include lighting and display technologies.

○ 这些功能依然要围绕艺术观念来决定。

The facilities of art museums are built upon art concepts.

○ 美术馆同样应该是绿色建筑。

Art museums should be green too.

○ 最大效益的建筑是使用得最充分的建筑。

The most efficient museum building is one that is thoroughly used.

○ 好用、节能的建筑设计不仅在于先进的技术设备，
而且也在于合理的设计概念策略。

A user-friendly and energy-efficient museum is the result of advanced equipment as well as reasonable design concept.

○ 应当避免制造问题再解决问题。

We should avoid producing problems, before seeking their solutions.

○ 减少对于人工的依赖。

We should reduce the reliance on manpower.

○ 合理的适用性标准：接上地气，合乎国情。

The standard of utility should be measured by local context and national conditions.

设计阶段需要考虑的博物馆专业设备

Museum Equipment Considered in the Stage of Preliminary Design

在这一节里，我们涉及的设备主要是那些在建筑设计阶段即需"嵌入"美术馆空间的设备。这些设备的尺寸、观感和使用往往对整个美术馆空间有着不可逆转的影响，早早考虑这些设备有利于避免补丁式的"二次装修"。

In this chapter, we highlight devices and equipment that must be "built in" during the preliminary design period of the museum projects. The size, visibility, and intended uses of these devices and equipment will have an irreversible impact on the museum space to be built. To consider them in advance will help to avoid patching-up renovation in the future.

- 照明和灯光设备
- 展示设备
- 储存设备
- 物流设备
- 安保设备

- Museum Lighting
- Display
- Storage
- Logistics
- Security

1/2 **室内环境控制**

需要独立出来的特殊的展览及保存环境，对湿度、温度等等要求在设计初期即需要考虑并予以规划。

Indoor Climate Control

The preliminary design needs to consider and plan for the special category of stand-alone display or conservation environment that needs temperature and moisture control.

3/4 **声光电网**

这些设备的分布和彼此关系对空间结构有着不可忽视的影响。展厅往往将它们集成在同一个基础设施系统中。

Audio/Visual/Power/Internet

The networks of audio, visual, power and internet have an impact on the configuration of museum spaces that can never be overlooked. A gallery often integrates them into the building infrastructure.

5/6 **储贮流通**

不同运输设备和贮存设备的尺寸预先决定了公共通道的尺寸。

Storage and Transportation

The size and volume requirements of storage units and transportation vehicles determine the dimensions of the museum's circulation spaces.

灵活使用和全时开放
Flexible Use and Full Operation

美术馆设计中应提高灵活配置空间的比重，特别是机动的展览空间要有效地补充总是捉襟见肘的永久展览空间。这样做的好处是带来新鲜而更有效的管理模式、各部分都能充分使用的美术馆空间、可移动的空间分隔、较高的效能和土地资源利用率。另一问题是，虽然大部分美术馆的日间开放时间是有限的，却通常占有城市的优势区位土地，地处繁华和中心区域，有必要考虑开放夜场，合理利用其公共空间面积。

Generally speaking, museum design should increase the quantity of spaces for flexible use. In particular, permanent exhibition spaces, of which there is a perpetual shortage, should be effectively complemented with temporary exhibition spaces. It will help to bring a fresh and effective management mode, fully utilized museum space, movable space partition, improved programmatic utility and land use efficiency. Most museums occupy primary urban sites in busy downtown areas while museum daytime operating hours are limited. It is necessary to consider maximizing properly the use of museum public spaces, for instance, by making the museum accessible at night.

1　芝加哥的菲尔德博物馆增加了晚间项目，方便下班后的人们参观，这样的老博物馆建筑在不同时间的开放需要对博物馆自身的安保进行重新评估和分级。

The Field Museum in Chicago operates evening time sessions for the convenience of visitors at after-work hours. In order to make such an old museum accessible to people at different times, it is necessary to reevaluate museum security measures, reworking approaches toward security grades and procedures.

2　亚利桑那州立大学的画廊直接面向大街，过路人可以看到画廊内装置的情况，类似一个"橱窗"。它对于夜晚城市空间的主要意义在于视觉上的透明性和心理上的共存感。

The gallery of Arizona State University Art Museum interfaces with passersby by a thin layer of glass, just like a shop window, that brings them "in" to the museum space from the street. It forms part of an integrated urban space through its visual transparency and psychological companionship.

24 小时美术馆
24-hour Museums

24 小时美术馆的概念，指美术馆的各部分可以在不同的时段运行而得到
不同的利用，甚至美术馆的主体部分闭馆之后，它的一部分空间还可以向
城市开放。即使没有借助高科技的节能设施，使用得最充分的美术馆已经
是最节能的美术馆。

The concept of the 24-hour museum means that the museum operates
and can be used, section by section, at different times. Even after the
main section of the museum is closed, certain parts remain open to the
city. Even without high-tech energy saving equipment, a museum that is
utilized at its maximum is an energy-efficient museum.

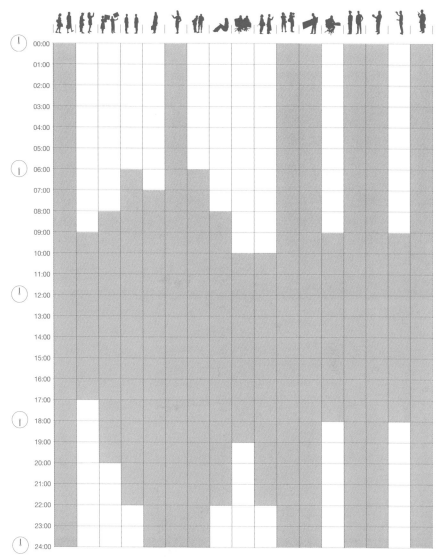

24 小时美术馆，建筑的一部分始终向观众开放。
A 24-hour museum has at least some parts open to the
public 24 hours a day.

绿色美术馆
Green Museum

绿色美术馆的设计首先涉及绿色建筑的一般原则：节水、节能和再生资源的使用。这里需要强调的是和美术馆自身功能特点有关的环保设计手法。传统的美术馆十分依赖人工照明和室内空调，重复的室内装修和展览搭建都是十分耗能或者浪费资源的。这些特点使得美术馆几乎不太可能成为环保设计的典型案例。

突破点或许在于美术馆自身观念的改变：部分开放式的美术馆可以有效地利用日光照明，把允许室外展示的展品放到半室外环境；采用分区空调方式提高能效；减少重复装修，使得展览设备可以反复使用；等等。

有条件的美术馆应该尽量和城市公共交通无缝结合。改善自身的周边生态也是营建绿色美术馆的重要手段。美术馆只有有限部分的展出空间需要严格的温度、湿度控制，其他部分的建筑设计应当尽量利用自然通风，利用景观绿化减少冬季的热能损失和夏季的日光暴晒。

Designing a green museum involves the adoption of general green building standards: water saving, energy saving, and reusable resources. We need to stress museum-related green design. Traditional museums rely heavily on artificial lighting and indoor air conditioning, as well as repetitive refurbishment and exhibition installation, which are energy consuming and resource consuming. These characteristics mean museums could almost never be successful cases of environment-friendly design.

A breakthrough may occur when the museum concept itself changes. Museums can be open to daylight so that art objects not sensitive to climate conditions can be installed in semi-open spaces. Smart zoning technology helps to reduce the need for central air-conditioning. Repetitive installing should be limited with reusable facilities. A museum should provide a smooth connection to the city's transportation system. Also, a museum should improve its ecological surroundings, for example, using natural venting when possible, and using plantation for temperature control in winter and summer.

两种不同的"绿色美术馆"观念，分别对应着建筑的生态环境和建筑使用的生态原则。

The "green museum" refers to one that is built in an ecological environment and one that is used according to ecological principles.

异型建筑
Irregular-Type Buildings

实践证明"异型"的空间在中国现阶段很难建造。不仅由于相对高的材料成本和设计费用，还因为在施工上有更高的要求。另一方面，相对于规整构造的建筑，"异型"的建筑的室内部分不易布置，利用率低且灵活性差。可以考虑适当减少异型空间和建造形式在美术馆中的比重。

It has demonstrated that constructing irregular-type buildings in China is no easy task, because they require relatively higher costs of building materials and higher design fees, and are more demanding in terms of construction quality. Furthermore, compared to regular-type buildings, buildings with irregular plans are difficult to furnish. Their interiors can hardly be used fully or flexibly. Museum designers should consider reducing the ratio of irregular-type spaces and building types.

结构相似的博物馆，构造了不同比重的异型空间，建设成本将大大不同。

Structurally-similar museums, depending on the ratio of irregular-type spaces, have drastically different construction costs.

....

标准型材和装配
Standard Building Materials and Installations

不同建构技术可选择使用不同种类数量的标准或定制构件。特殊定制构件无论建造和维护都较昂贵。标准构件通过优化设计可以部分或完全达到定制构件的效果。用"标准"产生"变化"要好过在繁复的"变化"中事后补求"标准"。博物馆建筑的装配方式尤其需考虑和展览设备的平滑对接。

Different building technologies may utilize different kinds and quantities of standardized or customized building components. Customized components are expensive to build and maintain. Standard components with optimized designs could partially or completely replace customized components. Creating varieties from the uniform is better than standardizing differences. It is especially important to consider the interface between the building and the museum equipment in order to ensure a smooth installation.

妹岛和世设计的玻璃艺术馆使用定制的曲面玻璃，尽管这些在中国定做的玻璃成本相对较低，但在运输的过程中损失了一部分。

The Glass Art Museum in Toledo, designed by SANAA, uses customized curved glass made in China. Despite its relatively low price, a portion of the glass were damaged during transportation.

本地施工
Local Construction

中国建筑材料的来源并不贫乏，合理开发利用本地特产的建筑材料将是减少项目预算的重要环节。与此同时，要合理地评估本地施工的工艺水平，确定相应的施工策略。能否在一两处具有本地特点的材料及其施工工艺上取得突破，往往是一个建筑项目是否能成功的关键。

There is a diverse selection of building materials in China. It is crucial to take advantage of local building materials and craftsmanship in order to reduce project budgets. Meanwhile, it is necessary to evaluate the proficiency of local construction teams and devise management strategies accordingly. The success of a project is made possible by one or two breakthroughs in utilizing local materials and construction craftsmanship.

上海世界博览会的西班牙馆在设计前即考虑了利用中国本地的藤编工艺。

Spanish Pavilion, Shanghai Expo 2010, utilized local craftsmanship to manufacture rattan work.

超高空间
Super-high Space

尽量减少高大的空间的比重。尽管现代空调技术可以对大空间的室内气候控制做分区处理，但是一个高大通敞的空调空间的能效总是较低的。在此产生的建筑维护成本将使建筑的有效运转难以为继。事实上，许多需要高大空间的艺术品都可以放在露天或半开放空间里。

It is desirable to reduce the ratio of super-high space. Although advanced air-conditioning technology can handle the climate control needs of spaces zone by zone, a super-high space that needs to be air-conditioned is always energy inefficient. Consequently, high maintenance costs can make a museum difficult to run. In fact, many art objects that require high vertical spaces can be placed in open- or semi-open environments.

高低不同的展览空间对空调系统的依赖不同。

Exhibition spaces of different heights have different degrees of reliance on air conditioning.

第九章

不同的视角
Different Perspectives

○ 美术馆不容易去，但可以在身边。
Art museums are not always available around, but can be made so.

○ 美术馆是为今天还是为未来的美术制造的？
Are art museums for today or for the future?

○ 好的美术馆里面的展品让人忘了美术馆本身。
Visible art objects lead to invisible art museums.

○ 如何用"光"是最重要的。
Lighting is of primary importance.

○ 美术馆与整个城市有隔膜。
The museum is an urban alcove.

○ 喜欢博物馆的窗。
I like museum windows.

○ 现在的艺术空间不只是一种收藏和展示的空间。
Contemporary art spaces are more than collection and display.

○ 美术馆应该是开放的，有趣可玩的。
Museums should be open and full of fun.

○ （使参观者）不累的美术馆。
A museum not tiring (for visitors).

○ 希望中国美术馆能寻找到展示中国艺术的方式。
Hope for a Chinese art museum for Chinese art.

注：本章访谈内容为简译。
Note: Translation for interviews in this chapter is simplified for reference.

王舒展 Wang Shuzhan	杂志主编 Editor
游旭东 You Xudong	杂志编辑 Editor
李兴钢 Li Xinggang	建筑师 Architect
周翊 Zhou Yi	美术馆艺术总监 Museum Director
伊尔卡·鲁比 Ilka Ruby	策展人和出版人 Curator and Publisher
汪民安 Wang Min'an	学者教授 Scholar and Professor
朱青生 Laozhu	教授 Scholar
王韬 Wang Tao	主编 Editor
魏浩波 Wei Haobo	建筑师 Architect
张云鹏 Zhang Yunpeng	学生 Student
Vivian Vivian	科技金融领域工作者 IT Worker in Finance
刘方 Liu Fang	退休中学语文老师 Retired Middle School Teacher
焦春华 Jiao Chunhua	图书编辑 Editor
张亚津 Zhang Yajin	规划师 Planner
魏皓严 Wei Haoyan	教授 Professor
冯江 Feng Jiang	教授 Professor
余斡寒 Yu Wohan	教授 Professor
何捷 He Jie	教授 Professor
石岗 Shi Gang	建筑师 Architect
刘克成 Liu Kecheng	教授 Professor
方晓风 Fang Xiaofeng	主编 Editor
王礴 Wang Sang	教授 Professor
朱晔 Zhu Ye	策展人 Curator

你自己设计 / 使用 / 参观美术馆空间时的最大体会？
What's your biggest concern when designing/using/visiting museums?

王舒展　美术馆不容易去。文化熏陶在大家生活里发生的频率应该更高；也许一个街道、小的院落或者广场这样的场所也可以承载美术馆的内容。

The museum is not always available around. Cultural cultivation should happen more frequently. A street, a yard or a public square can act as museum space.

游旭东　有些的美术馆给人的第一感觉非常压抑。

Some museums are, in the first impression, not uplifting.

李兴钢　（设计美术馆空间时）需要了解展品，需要了解以后这个空间里可能发生什么样的活动，然后才能开始设计。

(To design museums, we) need to understand the art object as well as what will happen in the space.

周翊　公营的美术馆，体制上的限制比较多；私营方面，运营者目前还没有意识到美术馆作为公共艺术空间的价值，考虑更多的还是成本的问题。

State-funded museums have certain institutional constraints while private museums still fail to realize the value of the museum as a public art space, giving greater consideration to the costs of maintenance.

朱青生　美术馆是为今天还是为未来的美术制造的？如果未来的美术我们不知道是什么，那么怎样给它留下实验的空间和试验的机会？"美术馆的第四功能"就是实验，是除了它的收藏、研究和展览教育的功能之外的功能。我们要留一个空间，把人类到目前为止没做过的事情和想做的事情在里面实验。

Is a museum built for today or tomorrow? If for tomorrow, how can we design space for future art if we don't know what it will be? "The fourth function of a museum" is experimenting. We should leave some room in a museum to experiment what we intend for the future.

王韬　我感觉好的美术馆里面的展品让我忘了美术馆本身建筑如何，印象最深就是哥本哈根美术馆看到的一组雕塑，让我彻底体验到了存在主义所描述的孤独感，其他啥都忘了。记得当年学建筑的时候印象最深的博物馆和美术馆，完全想不起来看到什么展品了。最近看的新美术馆，感觉艺术品都是拿来装饰烘托房子的，不知道是不是专业背景干扰欣赏艺术品的原因。

I feel that a good museum guides me seeing its art rather than the building itself, instead of the other way around.

魏浩波

如何用"光"是其最重要的考虑，我理解用"光"有三种层次：一是用于"看"的光；二是"制造氛围"的光；三是"光的组合拳"。

Lighting is of primary importance. I understand there are three levels: the lighting for seeing, the lighting for atmosphere, and the lighting as a combination of multiple functions.

Vivian

目前国内美术馆里的咖啡和餐点价格普遍偏高，不够大众化。

Most cafes and restaurants in museums aren't affordable for the general public.

刘方

我在体制内，传播主流意识的美术馆里参观者是被教育的对象。今日美术馆、时代美术馆、798这样民间的前卫的艺术空间给我的不仅仅是一种"受教育"的体验，还有启发。

As visitors I've learned a lot in public museums. In private, avant-garde type of museums such Today Art Museum and 798 galleries, I feel being not only educated but also inspired.

魏皓严

（参观时）太累……没完没了的作品，没完没了地看，没完没了地走啊，没完没了地欣赏……

(When visiting museums, I) feel so tired … endless art works, endless walking, endless viewing …

何捷

看美术馆半天一定虚脱，得吃东西、喝咖啡、恢复体力才能看下半场。罗马的Borghese Museum and Gallery规模刚好，而且楼上只能看一个钟头，去了两次都挺愉快。乌菲兹跟梵蒂冈博物馆都太大了。

After a half-day tour of a museum, one feels exhausted. You need to replenish your energy with eating and drinking.

张亚津 美术馆展馆和美术馆之间有隔膜，美术馆作品和日常生活有隔膜，美术馆与整个城市有隔膜。

There's estrangement between the gallery and the museum, between the museum's artworks and daily life, and between the museum and the whole city.

石岗 何香凝美术馆精致，作品全忘了，应该是建筑抢了戏。其他的美术馆主要是观看流线体验，又没有走重复路线。

I forgot about the artworks, and only saw the wonderful building of He Xiangning Art Museum. Other museums usually offer a linear experience and little repeated routes.

刘克成 美术如美人，美术馆如闺房。设计美术馆如找女朋友——怎么看怎么好，情人眼里出西施……参观美术馆如街头看姑娘，全凭心情。

Art is a like a great beauty. The museum is like her chamber. Design a museum is like seeking her as your lover.

朱晔 从策展人的角度希望美术馆好用。所谓好用，就是空间本身能够顺利地实现展览：能够适应架上、装置、影像、文本等不同类型的作品，墙面和天花板、地面都要好用，能够经得起适应性二次使用。同时，我也是个观众，所以不喜欢那种走半天闷在里面看展览的美术馆，看了一阵最好能够在身体和视觉上都有所休息。强制观展路线的美术馆，那是更糟糕的……

In the eyes of a curator, a good museum design is one that facilitates the process of exhibitions. As audience, I'd like to take random walks in a museum with spaces to rest visually and physically.

王碩 喜欢博物馆的窗，从冗长的展览的叙述中突然碰到一面有风景的窗真是十分惬意，是个视觉的休憩。印象中很多博物馆的窗都做得不错。如果像罗丹美术馆那样有个园子就更爽了，你知道在展览之后会有"犒劳"，有个地方可以舒服一下。

I like the windows of museums. It's very pleasant to gaze through a window to the outside while attending a tedious exhibition.

你希望中国未来有怎么样的美术馆？

What type of museums do you wish to see in China in the near future?

李兴钢　现在的艺术空间不只是一种收藏和展示的空间，现在美术馆的使用方式不再像以前那么单一、传统，而是有更多的公共活动、公共教育、对青少年的艺术熏陶，很多艺术活动本身就是在美术馆的空间里发生的。

A space truly for art and education, more than collecting and exhibiting.

周翊　提供更加充分的公共空间。目前大部分美术馆还比较传统，硬件上公共空间考虑得较少，只能通过使用上的软性方式来拓展公共空间。

More public space is needed. Currently, most museums are fairly traditional. Public space is given little physical considerations; furnishing is used as a remedy.

汪民安　美术馆应该是开放的，可玩有趣的，所有人都可以进去。没有什么界限，无墙的美术馆。任何事情都可以在美术馆里发生，各种奇思怪想、异端邪说都可以在美术馆里产生，它就像一个公共的思想广场一样。

A museum should be open and full of fun, open to everyone, without boundaries, without walls.

朱青生　我希望有一些实验的机会，让美术馆作为文化的汇合点、交融点来进行试验。我们现在所说的"艺术"都是已经有过的"艺术"而不是将要有的"艺术"。我们为何不把这样的一个汇合的中心放置在人们本来就在公共空间聚会的地方，反过来把美术馆空间扩大为人间的一次聚会。

Hope for a museum that provides more opportunities for experimenting, a museum as the confluence and intersection of cultures.

魏浩波　传统的美术馆做法是通过生产某种特定的固有场所供养美术品，即便是杜尚把尿壶扔进了美术馆，可美术馆依旧是圣殿，反倒是那尿壶鸡犬升天似的，这是"真空"般形而上方式；而中国是泛普罗大众的社会，美术馆做成流动的货柜车，上山下乡，更有可能成就普及文化的作用吧。

We need a "museum truck" that could go anywhere to popularize the culture it contains.

Vivian	更加生活化、常态化的美术馆，而不是百姓眼中"高大上"的殿堂。 A museum that is closer to daily life, instead of a temple.
焦春华	美术馆最好能走进社区，在家门口就能看到很棒的展览。 A museum in the community that allow you to see good exhibitions in your neighborhood.
魏皓严	（使参观者）不累的美术馆，可以躺着坐着趴着看，也可以不看，作品只是风景中被守护的酱油客，被游览的人无意中遇见，或者不必遇见。 A relaxing museum, one in which you could sit or lie down to just watch. Or just ignore the art exhibits.
张亚津	坐在留园的长案前，八扇落地长窗推开，窗外一池新荷。长案上，并肩的人徐徐打开一册手卷。 At the Lingering Garden, viewing a scroll of painting at a long table in front of eight screen windows.
石岗	……我和你多年交往后，互相了解，你在我的或者我在你的作品前，我们放松地相互介绍和解说，或者漫谈一下，那时的那种空间是最好的，比如饭桌前、乒乓球台前、去机场的大巴里、湖边、你家、我家，其他业务合作偶遇时，一起逛城中村时…… A space where you and I meet (after years of friendship) in front of your or my artworks, where we can share intimate and casual conversation.
刘克成	希望中国美术馆能寻找到展示中国艺术的方式，不要只跟着西方人的屁股后面跑。中国传统绘画属于情境艺术，就不完全合适放在西式画廊里面。 A Chinese museum for Chinese art that does not blindly follow Western conventions. Chinese traditional painting is a contextual art, which does not fit in Western-style galleries.
朱晔	我希望中国未来有更加有趣的美术馆，但是更希望中国没有美术馆，它们应该在大街上、在菜场、在居民区、在工厂旁边、在农村、在鱼塘…… A museum that is more fun. Or no museums at all. They should be in the street, in the market, in the community, by the factory, in the countryside, at the fishpond...

你见过的世界或国内最好的美术馆和展览空间？

What is your favorite museum and exhibition space in China or in the world?

王舒展　国外的很多美术馆空间很棒，令人印象很深刻，但它们和我个人的人生体验关联度比较低，没有和我个人的情感发生关联和共鸣……身边的美术馆很少能在一个人内心深处真正打动人。

Some foreign museums are good and impressive, but they are not quite relevant to my life... while museums around me rarely touch my heart.

游旭东　北京的红砖美术馆。我觉得它作为艺术的容器很有意思，美术馆本身就是一个展品，红砖看起来很亲切。

The Red Brick Art Museum in Beijing is very interesting as a container of art. The museum is an exhibition item itself. The red bricks look amiable.

李兴钢　路易斯·康设计的美国的金贝尔美术馆。

Kimbell Art Museum designed by Louis Kahn.

周翊　尤伦斯美术馆。硬件方面，建筑质量相对好一些，位置上也处于一个艺术氛围浓厚的地带。软件方面，开放的公共交流项目比较多。

UCCA Center for Contemporary Art. Good in location, building quality, and communication.

伊尔卡·鲁比　这并不能一概而论，因为美术馆空间的好坏还与它即将展出的东西和方式有关。我最喜欢去的美术馆是巴黎的东京宫（Palais de Tokyo），它不是一个白盒子，原先这座建筑是为了别的目的兴建的，现在它已然成为一个影院，并且建筑物的一半已经拆除了。我喜欢这样的方式：艺术是中立的，是在讲述着什么的空间中展示出来。

Palais de Tokyo in Paris. A good museum cannot be assessed without understanding of the specific context of its exhibitions. For me, art is neutral, and should be displayed in a narrative space.

汪民安　相对而言，798 的尤伦斯美术馆、中央美术学院美术馆做得比较好。

Relatively speaking, UCCA Center for Contemporary Art in 798 Art District and CAFA Art Museum.

朱青生　蓬皮杜艺术中心，卢浮宫，中国的故宫，让·努维尔设计的巴黎的 Musée du quai Branly，古根海姆美术馆。

Le Centre Pompidou, the Louvre, the Palace Museum, Musée du quai Branly, the Guggenheim.

魏浩波	德·莫拉的保拉·蕾格博物馆，以一组粗狂的红色混凝土罩面的不同高度、不同大小的盒子群定义不同的功能展厅，各展厅盒子以对角开口形成连续的有张力的空间序列，成功地协调了单元盒子的独立性与整体盒子群连贯性的矛盾，解决了同一单元盒子内穿越路径与静态观看区的相互干扰问题，可谓一招化解式。
	Paula Rego Museum in Portugal by Souto de Moura.
张云鹏	国家博物馆。
	National Museum of China.
余斡寒	乌菲兹美术馆。
	Uffizi Gallery.
张亚津	蓬皮杜结构清晰，酷劲儿十足，交通集约，展览空间感觉巨大；还让了一半场地给城市广场，设计极简，是我印象中巴黎最有活力的广场。巴塞罗那现代艺术中心，迈耶的设计，有更丰富的公共空间品质，可惜最后剩下一点点展览空间，喧宾夺主。
	Le Centre Pompidou. Contemporary Art Center in Barcelona by Richard Meier.
方晓风	波士顿美术馆边上的伊莎贝拉·加德纳博物馆……特点是建筑把藏品集成在一起，中庭又布置成四季花园，很有女性气质……在其中参观像去她家做客，轻松而舒服。其扩展部分是皮亚诺的设计，特意做了可以休息的客厅，亲切。
	Isabella Stewart Gardner Museum in Boston.
朱晔	……木马吧，蔡铿开的小酒吧，时不时做小展览……就是日常的消费空间。
	Daily spaces with small-scale art shows, like the Trojan Horse Bar.
王硕	蓬皮杜，因为它的广场实在是很有吸引力，走过路过都愿意绕一脚从那儿坐一下，有的时候是为了去这个地方才顺便去看个展览。城市公共空间与博物馆以及图书馆相依相生，才形成了如此重要的蓬皮杜。
	Le Centre Pompidou, especially the plaza in front of it. The combination of its public space, museum, and library makes it attractive.

第十章

行动指南
Guidelines for Action

○ 美术馆设计需要深入细致的调研和清晰系统的规划。
Museum design demands in-depth and careful research as well as clear and systematic planning.

○ 想好才能做好。
Careful consideration should be initiated before action.

○ 没人可以代替业主自己对美术馆的了解和定位。
No one can replace the client himself or herself in terms of understanding and positioning of a museum.

○ 完整的建筑策划过程开始于对大问题的思索。
A complete museum plan starts with big questions.

○ 多方合作，切磋出完整的功能程序。
Multi-side collaboration makes a full museum program.

○ 明确工作量和工作方式，紧控关键工作节点。
It is necessary to specify the work load and the approach to the work in order to control the process and timing.

○ 必要的工作模型。
A working model is necessary.

○ 考虑长期改变和扩展的可能。
It is necessary to consider the possibility of change and expansion to the museum in the long term.

○ 有相似的美术馆空间，没有两座相同的美术馆。
Museum spaces might be similar, but never being identical.

○ 美术馆最终应当是属于艺术的。
In the end, the art museum belongs to art.

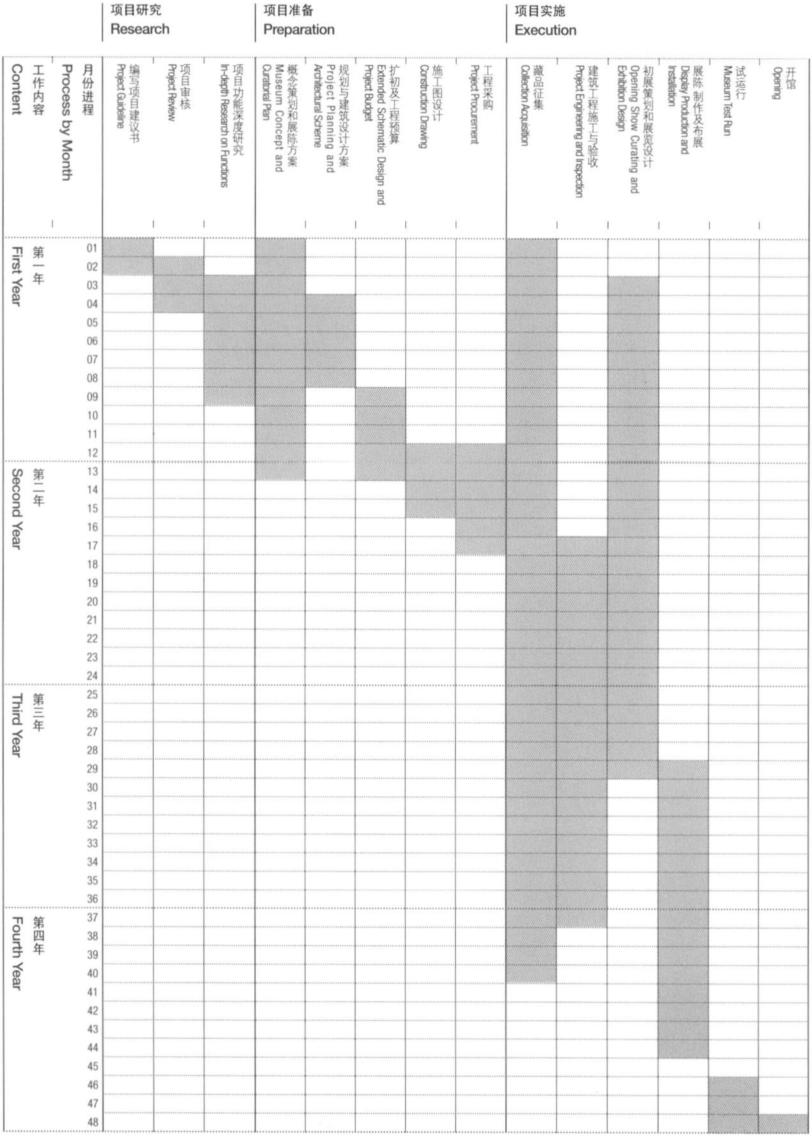

项目研究
Research
项目准备
Preparation
项目实施
Execution
工作内容
Content
月份进程
Process by Month
编写项目建议书
Project Guideline
项目审核
Project Review
项目功能深度研究
In-depth Research on Functions
概念策划和展陈方案
Museum Concept and Curatorial Plan
规划与建筑设计方案
Project Planning and Architectural Scheme
扩初及工程预算
Extended Schematic Design and Project Budget
施工图设计
Construction Drawing
工程采购
Project Procurement
藏品征集
Collection Acquisition
建筑工程施工与验收
Project Engineering and Inspection
初展策划和展览设计
Opening Show Curating and Exhibition Design
展陈制作及布展
Display Production and Installation
试运行
Museum Test Run
开馆
Opening
第一年
First Year
第二年
Second Year
第三年
Third Year
第四年
Fourth Year
01
02
03
04
05
06
07
08
09
10
11
12
13
14
15
16
17
18
19
20
21
22
23
24
25
26
27
28
29
30
31
32
33
34
35
36
37
38
39
40
41
42
43
44
45
46
47
48

美术馆项目进程参考

Typical Timetable for Museum Projects

本表格对美术馆项目，尤其是公立美术馆项目的筹备建设提供了时间节点上的参考。工作次序和工作内容的清晰有利于项目管理，但值得指出的是，实际工作中的难点突破往往在于各项内容的彼此协同，而不全是各自为战和按部就班的。

This table provides a reference for the timeframe of a museum preparation project, especially a public one. Clear working order and task requirement are helpful for project management. However, the difficulty in practice often comes from the coordination of each task. Isolated work mode and blindly following should be discouraged.

美术馆项目四部曲
Tetralogy of Museum Projects

1

美术馆项目筹备
Museum Project Preparation

和建筑设计直接相关的美术馆项目筹备工作，包括资源和管理上的两大块。尤其重要的是美术馆的收藏从哪儿来？将来的运营方式是什么样的？这方面的决策将会直接影响到美术馆的设计。

Preparatory tasks that are directly related to architectural design include collection resources and museum management. Most importantly, where does the museum's art collection come from? How does the museum operate? Decisions concerning both of these questions have an immediate influence on the museum's design.

2

美术馆项目设计管理
Museum Project Management

适时地分阶段引入一个有意思的美术馆设计理念，而不一定"毕全功于一役"。有趣的、吸引各方注意的设计将会推动美术馆筹建，美术馆项目自身的进展也会倒过来给设计提供具体的要求。

Project management starts from introducing innovative concepts of museum design at an appropriate time and in different phases. It is not necessary to decide everything at one time. Interesting and attractive design ideas can drive the process of museum preparation. Conversely, the progress of museum preparation will provide more specific directions to museum design.

3

典型美术馆空间设计
Typical Museum Space Design

美术馆的设计一定需要从一开始就要考虑细节和最终完成的状态，有时候需要建立一个实物大小的"模型"；它用起来会是什么样的？建筑设计全部结束再考虑装修和空间分配，往往会造成顶层设计和人际感受的撕裂。

Museum designs will have to, in the first place, consider various details as well as the final look. A one-to-one model may be necessary in order to check the actual conditions of use. If we consider interior decoration and space configuration after the museum is completely built, a gap between top-down design and the actual use at the personal level may emerge as a result.

4

有创意的和为明天的美术馆
A Creative Museum for the Future

很多美术馆还未完工就已经过时，有的美术馆中规中矩却不人气寥寥。好的美术馆设计需要给未来的发展留下余地，不妨建立一二实验性的、灵活使用的空间。

Some museums are already out of fashion before their completion. Some museums have mediocre designs and thus are not frequented by audiences. A good museum design will leave enough room for future development and will allocate a few experimental spaces that can be flexibly used.

建筑招投标和设计任务书
Architectural Bidding and Design Guidelines

建筑招投标和建筑竞赛的利弊

为了寻求新鲜的建筑设计概念,重要项目通常会采取建筑招投标和建筑竞赛的方式。建筑招投标和建筑竞赛的本质特点是个人之间的竞争,它有利于产生出对同一问题多样化的解决方案和鼓励创新。一部分此类项目的评议采取盲评的办法,也激励着参赛建筑师,无论有名与否,拿出他们最为精彩的设计。

但是,这种方式也有很多弊端。首先,竞赛本身可能有很多人为因素干扰导致竞赛无效,例如,1976、1979 和 1989 年,希腊政府举办了 3 次建筑竞赛来决定谁设计用来陈列雅典卫城文物的卫城博物馆,但是都宣告无效,直到 20 年后该建筑才终于宣告建成;其次,通过个人竞赛方式产生的方案很难做到周全的考虑,业主和建筑师之间的关系为有关法律约定而可能缺乏彼此信任,有时候中标的建筑师未必具有圆满实施方案的后续能力。

Architectural Bidding and its benefits and limitations

In order to seek fresh concepts, important projects often hold architectural bids or competitions, which by their nature are competitions among individual designers and their personal ideas. It is good at finding diversified and innovative solutions to raised questions. Some competitions review submissions in anonymous mode. In this way, ideas are prioritized over fame and notoriety, giving incentives to architects for best ideas.

But such practices have many problems. First, there are countless human factors that might cause the competition to fail. For example, in 1976, 1979, and 1989 the Greek government held three competitions for a new Acropolis Museum, but they all failed to produce an actual plan. The museum was only built twenty years later. Second, design schemes based on personal competitions often fail to take comprehensive consideration of the entire situation. The relationship between the client and the architect is governed by various regulatory laws, and this may result in a lack of trust between the two parties. Sometimes the winning bid design cannot be realized in practice.

建筑设计任务书

最基本的建筑设计任务书只包括业主对建筑未来功能的简单预设。一本更详尽的任务书需要定性定量地给出功能的结构组成，对于创新和亮点的期待，甚至举出具体例证，但是建筑设计任务书应该遵循"模糊决策"的原则，它不应限制设计者的思路，而只是深入对话的框架与平台。

Guidelines for Museum Design

The basic version of Design Guidelines articulates the owner's simple expectation of building functions. A detailed version of Guidelines specifies functional structure of the building in qualitative and quantitative measures. To achieve a creative and attractive design, the Guidelines often cite examples. It is advised that the Design Guidelines should be "vague" in principle, so as not to restrict architects' ideas and approaches, but to serve as a platform for dialogue and communication.

建筑方案的内容组成

The Components of Architectural Design Schemes

建筑方案的内容和编制深度的设计说明书需完整清晰地简述美术馆设计的特点，并从各个方面分别说明建筑设计的创意和思想。这其中尤其需要突出的是与美术馆主要专业功能，即艺术展览所对应的设计手法。同时，美术馆设计中所涉及的公共空间设计要与建筑的城市策略大致对应。最后，设计方案还要对所涉及的主要技术手段和特色进行说明。

一方面是建筑师被动地适应主要经济技术指标的硬性规定；另一方面，建筑师也需要主动地从各个层次分析主要建筑创意的来源和实施可能。这就要求建筑师必须提出一套完整的设计图纸而不仅仅是建筑的"效果"，业主应该了解建筑设计的基本特点以及各图纸所反映的设计内容。建筑设计方案最初的图纸不应一概要求"专业性"，以便于设计各方的沟通为好。

设计成果的形式需要多样化，从直观的多媒体演示文件到标准设计文册（通常为 A3 版面）、图纸展板，以及以上内容的电子文件，必要的时候，1∶1 的建筑片段实物模型（大样）有利于业主和建筑师实际评估建筑设计的可行性。

The design scheme document needs to succinctly specify major design features of the museum and describe the innovative ideas and thoughts from different perspectives, esp. regarding the primary function of the museum (e.g. art exhibition), as well as public space design conforming to urban strategy, and technical methods and features.

On the one hand, architects need to adapt their designs to the nonnegotiable standards of major economic and technological indexes. On the other hand, architects need to, more proactively, analyze the project regarding its resources and potentials at all levels. This requires that architects provide a whole set of design documents instead of merely a "final blueprint". The client should understand the essentials of architectural designs and the content of documents. In architectural design scheme, the technical specifications should facilitate communication between the various parties, instead of being overemphasized;

The design documents need to be diversified regarding format, ranging from multimedia displays to standardized design booklets (in A3), display boards, as well as electronic files thereof. When necessary, one-to-one architectural mockups or models will help the client and architects evaluate the feasibility of architectural designs.

中国国家美术馆新馆竞标方案
Competition Submissions for the proposed New Building of National Art Museum of China

长时段设计管理
Long-term Design Management

096 卢浮宫博物馆扩建 The Louvre Expansion

对于业主和设计师而言，面面俱到的专业咨询也许是不必要的，但是一定要对本书前七章提到的那些大类问题做出充分的思考。这其中城市的问题取决于"此时、此地"，而艺术展示问题则取决于美术馆本身的功能规划，包括展品、展览和展事。

完整的功能分析和基本推导是非常必要的。类似"实战演习"那样的功能结构地图有助于了解美术馆设计中出现的失误和缺陷，鉴于功能结构地图不可能覆盖所有使用者的兴趣和使用方式，应该类似本书第九章那样，变换视角对于功能结构地图做出新的调整。

设计导则的意义在于更具体地给出业主对于建筑设计方向的意见。这些意见的目的不是为了在形态上限制建筑师的思路，而是对那些项目中一些特殊的兴趣点做出符合功能的设计导引。设计导则需要结合装修、设备等对于典型空间的样式、材质和使用做出具体的引导。

一个设计是否能够完满实现取决于很多现实条件的配合：施工工作的开展、特定建材的有无、配套设备的好坏、工程监理的完善都是需要考虑的对象。应当在设计环节就避免那些难以实现的构想和要求。

一个完善的设计要综合长期因素和短期因素，考虑到未来变化和扩展的可能，这种可能既包括物理规模的增加也包括因为新技术带来的美术馆运营模式的改变。前者需要考虑建筑的预留用地，而后者需要考虑管线和设备未来更新的要求。

大都会艺术博物馆今昔对比

大都会艺术博物馆的扩建用地来自于它背后的中央公园，作为寸土寸金的城市，纽约慷慨地允许大都会博物馆在公园用地中占据可观的一部分空间。扩建的部分在形式上呼应了公园的自然条件，形成了若干可以"呼吸"的天井和平滑的过渡。

The land for the expansion of the Metropolitan Museum of Art came from Central Park behind the museum. In the city where land is so valuable, New York generously allocated a considerable portion of park land to the museum. In return, the museum expansion was designed to echo the natural conditions of the park, forming several "breathable" courtyards as well as smooth transitions between the new and the old, the urban and the natural.

Clients and architects need not make an encyclopedic inquiry into all issues but should thoroughly consider big questions raised in the previous seven chapters, especially urban issues, determined by "here and now" and art exhibition issues, determined by functional planning (including exhibits, exhibitions, and events).

It is necessary to perform function analysis and simulation. A complete structural map of museum programs, like a "rehearsal", help to understand possible errors and failures in museum design. As such a structural map cannot possibly accommodate the interests and needs of all museum users, it can be adjusted based on different perspectives (as in Chapter 9).

Design guidelines help the clients to communicate their expectations of museum design. Their opinions are not meant to restrict architects, but to guide the project in terms of specific needs and functions. Design guidelines also need to give detailed directions on the type, materials, and function of a typical museum space, in connection with furnishing and equipment.

Whether or not a design is successfully realized depends on many practical factors: construction work's quality, availability of specific building materials, quality of museum equipment, and quality of construction inspection. A proper design should anticipate and avoid impractical demands.

A good design needs to consider long-term and short-term use and future possibilities for change and expansion. This includes both the physical transformation and the upgrading of museum operational model with technological advancement. The former concerns the reservation of enough land for future uses while the latter concerns upgrading of infrastructure and equipment.

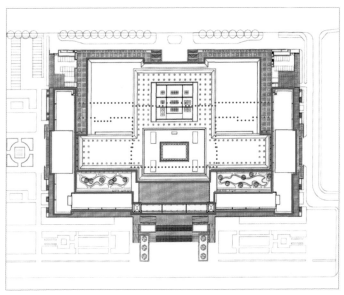

中国国家博物馆（改扩建）实施方案总平面图
Ground plan of expansion of National Museum of China

N 0 20 60m

中国国家博物馆改扩建前原状态
Before

中国国家博物馆在 21 世纪初的扩建用地来自于原建筑的天井面积和东侧公安部让出的空间，原来的露天开放空间的一部分被转化为一个整块的高大室内空间。

中国国家博物馆改扩建后
After

The National Museum of China expanded at the 21st century in its original courtyard and in an area given by the Ministry of Public Security to its east. The original courtyard is partially converted into a lofty interior space.

典型公共空间的设计

The Design of Typical Public Spaces in Museums

公共空间应该尽量利用自然光照明，和室外空间、露天景观适当结合；大部分美术馆的公共空间应适用人际尺度，和火车站、旅馆、机场等过于嘈杂的公共空间拉开距离。

无论多么富有创造性，"艺术"依然是公共空间的主角，围绕着核心功能而"混合"设计的公共空间，可以生发出肃穆的展厅或热闹街区所不具备的精神氛围。

Public spaces in museums should utilize natural lighting and connect with outdoor spaces and landscape. Most public spaces in museums ought to be human-scaled, different from railway station waiting rooms, hotel lobbies, or airport lounges.

No matter how creative they are, public spaces should still give highlight to "art". With art as their core function, these public spaces can develop a unique "mixed" feel that neither a solemn gallery nor a noisy street could achieve.

1　美术馆的花园：纽约现代美术馆洛克菲勒雕塑花园
　　Garden of an art museum: the Abby Aldrich Rockefeller Sculpture Garden, MoMA, New York City
2　花园中的美术馆：洛杉矶盖蒂中心
　　Art museum in a garden: the Getty Center, Los Angeles

典型画廊空间的设计
Typical Gallery Design

美术馆不是从来如此的，艺术展示的问题也没有"必然如此"的定论。艺术和艺术理论的传统，比美术馆建筑学的历史要长得多，它们提出了一些关于艺术展示的经典问题。

● 画廊空间设计的核心是艺术展览，它不仅仅限于"看"的问题，当代艺术展览尤其如此。当代美术馆展厅的设计不限于一般的"画廊"模式，无论"黑盒子"，还是"白盒子"，都只是可能性中的一种。

● 艺术展览中的"看"，不仅仅是关于"看什么"，也是关于"怎样看"的问题。当代美术馆展厅的设计要结合"静观"和"动观"，要深入地研究现有展品和展览，并结合美术馆的当代公共空间品质。

● 关于"怎样看"，美术馆展厅的设计需要考虑展览的私人或公共氛围，艺术品和观众的尺度关系、展厅和展品的照明方式，以及展厅空间结构中所预设的展陈结构。

There is no fixed pattern for museums, nor is there a fixed rule on how artworks should be displayed. Art and art theory boast a longer history than art museum architecture, and have raised some classic questions concerning art display.

● The design of a gallery concerns the core issue of art display. It should not only focus on visitors' "vision"; this is especially true with contemporary art exhibitions. The design of art galleries should go beyond the conventional "gallery" model, where the "white cube" or the "black cube" presents only ONE possible model, among many alternatives.

● "Viewing" art exhibitions is not only about what to view, but also how to view. To design an art gallery of this age, architects should explore and gain a deep understanding of the exhibitions and current objects on display, and take into account the character of the museum as a public space when trying to create a combined "reflective" and "interactive" experience for visitors.

● A good art gallery design presumes a thorough consideration of the following factors of "viewing": the show experience, private or public; artwork scale in proportion to audience; lighting solutions for gallery space and artwork; and the preset display structure in museum spatial setting.

菲利普·约翰逊在自己康涅狄格州新迦南的别墅中设计了几类不同的画廊空间。
Philip Johnson designed several gallery spaces in different types in his
Glass House in New Canaan, Connecticut.

封闭的室内空间展出 2D 作品，活动展墙可以绕轴心旋转。
Two-dimensional artworks displayed in an enclosed exhibition
space. Walls can be turned around the axis.

开放式自然光照明的雕塑展厅，有着透明的天顶和起伏的室内地形。
The Sculptural Gallery is naturally lit with a transparent ceiling and interior topography.

典型的西方画廊模式
Gallery: The Typical Western Model

从文艺复兴以来,典型的博物馆展出仅仅是把展品分为二维和三维的展品,这些展品和人之间是界限分明的客体和主体的关系。展品的定义清楚,构造明白,展览的目的就是营造中性的环境突出展品,人和展品"看与被看"的关系也比较简单。

- 将展品和它的环境清楚地区别开来,同时也清楚地区分了"主体"和"客体"之间的界限和距离;
- 把展览对象看作性质类似的"物品",对它们的来源和"原境"不加区分,同时,展览的环境中性化,所有的展厅设计和展品照明设计一视同仁。

Since the Renaissance, a typical museum exhibition differentiates its objects only by two-dimensional and three-dimensional types. In both cases, exhibits to audiences are like object to subject. The exhibits are clearly defined and structured. The purpose of the exhibition is to show the exhibits in a neutral environment. The relationship between the exhibits and the audience is simple and clear as "to-be-seen and to-see".

- When the objects are separated from their environment, clear boundaries distance are set between "subjects" and "objects".
- Works of art on display are treated as similar "objects", without differentiating their sources or their original context. The exhibition space is meant to appear neutral, with all the exhibition halls and lighting schemes treated the same way.

1　巴塞罗那1929世界博览会德国馆,密斯·凡·德·罗设计。空间本身是一件艺术品,但是依然有其"展品"。

The German Pavilion, designed by Mies van der Rohe at the International Exhibition held in Barcelona 1929. The space itself is a piece of artwork even though the space displays its own "artwork".

2　"新"和"旧"的对比。

Contrast between the old and the new.

传统式样中国空间的艺术创作和展示
Art Creation and Display in Traditional Chinese Spaces

虽然中国大多数当代美术馆的展览重心是 20 世纪以来的艺术，但是传统式样的中国空间对艺术创作和展示的影响依然体现在方方面面。在向西方学习的初期，大多数展示空间全盘照搬了西方美术馆的画廊样式，但是未来这些美术馆的设计需要不限于典型的"白盒子"模式，而有更多的中国特色。

故宫博物院是当代中国少有的、利用传统中国建筑作为展厅的艺术展示空间，这种艺术展示空间中同时蕴含着矛盾和希望。传统的中国建筑和中国艺术固有的观赏模式是贴近的，但是这种建筑作为当代展厅的利用率较低。除了建筑本身造型复杂、难以将就，还涉及传统艺术品的私人欣赏氛围是否适用于当代的公共社会。

Although Chinese contemporary museums focus on art from the 20th century onwards, the influence of traditional Chinese spaces on art creation and display can still be felt in all aspects. At the early stages of learning from the West, most exhibition space designs copied the gallery model from Western museums. In today's context, Chinese art museums need to go beyond the constraints of the typical "white box" gallery model, and should infuse the spaces with more Chinese elements.

The Palace Museum is a rare example of utilizing a traditional Chinese building as exhibition space. This kind of exhibition space embodies both contradiction and hope. Traditional Chinese buildings and tradition art feature a close-up view mode. However, traditional buildings do not provide enough exhibition capacity in modern context. In addition, the complicated forms of spaces in traditional buildings are not ideal for configuration. On top of that, it is still to be decided if the traditional close-up view mode in private exhibition context serves the modern public well.

现代时期重新设计改建的武英殿展厅
Wuying Hall Gallery remolded in modern times

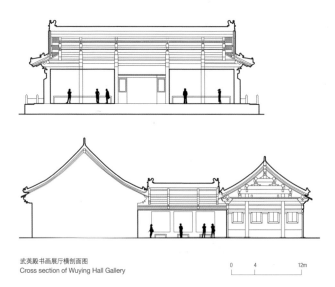

武英殿书画展厅横剖面图
Cross section of Wuying Hall Gallery

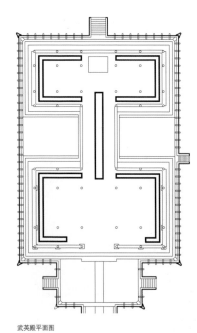

武英殿平面图
Plan of Wuying Hall Gallery

李兴钢设计的绩溪博物馆，在传统建筑类型中引入自然光。
Jixi Museum by Li Xinggang. Natural light is introduced into a traditional building.

苏州博物馆临时展厅：贝聿铭的设计理念是"中而新，苏而新"。
Gallery for Temporary Shows at Suzhou Museum, by I. M. Pei. His design motto is: "Chinese but new, Suzhou but new."

展览空间的灵活性研究

Research on Flexible Display Spaces

展览空间的灵活性议题包括两个部分：首先，一个美术馆将需要多样化的展厅，包括不同的尺寸和形式；其次，各种展厅间应有一定的互换性，才能确保将来同时满足各种需求。总体上，不同展厅的组合达到的是展览空间使用的最优效益，并不单纯在于"大"和"多"。

最大化展厅提供了可供将来灵活分割的整块空间，墙和出口按一定比例关系和相对位置的组合，决定了可变而多样的展陈方式，同时形成一定规律的布展模式，让展厅的容量变得更易预期。自由布局和特色展厅补充了过于规则的展厅在展览潜力上的不足。

The flexibility of gallery spaces implies two aspects: first, an art museum needs diversified galleries of different sizes and layouts; second, galleries should be interchangeable to some degree in order to satisfy different needs. In general, the combinations of different galleries optimize the uses of museum spaces. They do not need to be large in scale or quantity only.

A gallery with maximized capacity provides the flexibility of dividing spaces further when needed. By clever partitioning, walls and outlets are positioned to create multiple options and patterns for exhibitions. The different patterns of exhibition installation developed in such a way makes it easier to measure spatial capacity. In addition, galleries of free plans and customized features complement regular-shaped galleries with more potentials for display.

展厅尺寸　最大可能的单一展厅

Size of Gallery　A Gallery of Maximized Capacity

金泽 21 世纪当代美术馆剖面图
Section of 21st Century Museum of Contemporary Art, Kanazawa

0　5　15m

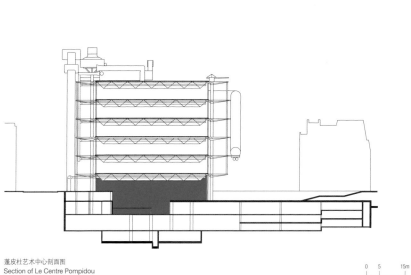

蓬皮杜艺术中心剖面图
Section of Le Centre Pompidou

0　5　15m

展厅序列　　不同大小和规模的展厅
Gallery Series　　Galleries of different sizes and scales

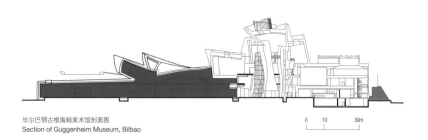

毕尔巴鄂古根海姆美术馆剖面图
Section of Guggenheim Museum, Bilbao

0　　10　　　30m

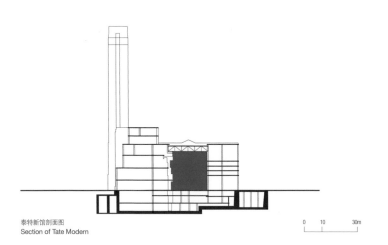

泰特新馆剖面图
Section of Tate Modern

0　　10　　　30m

展厅效益
Effective Galleries

展线、展品、和空间与体积关系的研究
Study of Circulation Route, Art Object and
Space-Volume Relationship

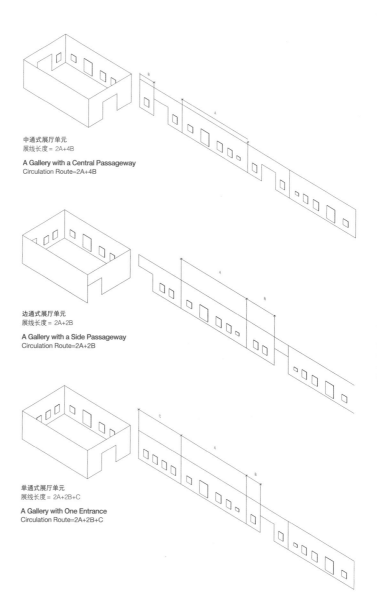

中通式展厅单元
展线长度 = 2A+4B

A Gallery with a Central Passageway
Circulation Route=2A+4B

边通式展厅单元
展线长度 = 2A+2B

A Gallery with a Side Passageway
Circulation Route=2A+2B

单通式展厅单元
展线长度 = 2A+2B+C

A Gallery with One Entrance
Circulation Route=2A+2B+C

照明设计
Museum Lighting Design

"光"是美术馆设计的灵魂，传统的美术馆照明完全或大部分依赖人工照明，将照明层次区分为局部、区域和环境，力求"浑然一体"。随着艺术本身的变化，美术馆越来越多地引入自然光照明，并且将照明效果看作美术馆设计的一部分，倾向于使其戏剧化而不是掩饰它的存在。

Light is the soul of museum design. Traditional museum lighting completely or mostly relies on artificial lighting. Lighting needs can be categorized into local, regional, and environmental; designers should strive to create a space of light as a whole. With evolution of art, art museums have introduced more and more natural lighting. Being considered a part of museum design, museum lighting tends to be dramatized instead of being downplayed.

伦佐·皮亚诺设计的洛杉矶郡美术馆，照明设计直接影响建筑的外观。

Los Angeles County Museum by Renzo Piano. The lighting strategy directly highlights the building's appearance.

路易斯·康的金贝尔美术馆，在貌似简单的
建筑类型里嵌入的照明系统，将自然光和人
工照明整合在类似的外观里。

Kimbell Art Museum by Louis Kahn.
Here, the lighting system is subtly
embedded in a seemingly simplistic
building that integrates natural and
artificial light.

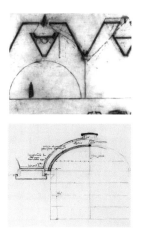

顶光
上方的漫射光具有神圣的气氛。

Top-lit
A sacred and religious atmosphere,
created with light diffusion from above.

逆光
勾勒出三维造型的轮廓。

Back-lit
Three-dimensional profiles enhanced by
backlighting.

界面
在不甚透明的界面两侧有着不同的观感。

Interface
Different senses on two sides of an opaque
interface.

透射
自然光照明成为室内空间的主调。

Lit-through
Natural light becomes the main lighting of
the interior space.

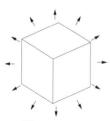

自明
展览空间的照明同时影响到邻近。

Self-lit
A self-lit space becomes the lighting
source for its environment.

结语

Afterword

一点哲学

唐克扬

美术馆设计需要一本什么样的"设计手册"？这本书不是一般意义上的"建筑技术速查手册"。它的首要任务是明示、分解乃至厘清美术馆设计需要面对的议题——因为，我们在实践过程中确实感受到，一个项目的**起点**有时甚至比它的解决之道更加重要。为什么要建美术馆？美术馆到底有什么用？怎样使一座美术馆的功用在当下中国的情境里最大化？这些都是美术馆建筑设计启动前就要考虑的问题。

美术馆设计不是一个纯技术活，它的文化前提难以一概而论，但是又不得不寻求某种共识和依据。因此，这本书一方面希望尽可能地打开思路，另一方面又尽量避免太过"个性"的话题。它确立了四个由大到小的思路：**城市、艺术、建筑**设计和**技术**。这里的顺序并非褒此贬彼，既然实践领域中尚没有一种可靠的理论和准则，我们只好着眼"起点"的问题，先在读者中普及某些重要的话题，并不奢望以一个答案达到世界一流水平。

但是本书并不是没有自己的观点，它源自于我们工作对象的特殊性：艺术。艺术是什么？和人类文化的其他门类不一样，当代的艺术实践不是试图一劳永逸地"解决"某个问题而是着重"打开"**很多**问题。在过去 200 年内一直在持续发展，现代意义上的"中国艺术"其实并没有"定案"，中国当代的美术馆设计，也得为这尚未落定的艺术实践留下点面对未来的空间。如今美术馆的设计多是讨论外部形象和内部风格，建筑师就算偶有"创意"，也挤不出可以深谈的道理。如此地准备不足，我们却迎来了美术馆的"大跃进"……**建筑**不理解艺术，**艺术**不满意建筑，**城市**有时对艺术、建筑两者都不"感冒"，更不用说很多**技术**的硬伤——风险和遗憾可想而知。

本书的写作编辑正是针对这样的语境。它不再是百科全书式甲乙丙丁、1234 的流水账，它试图确立更合理的**结构**和**层级**。首先我们需要考虑当下的中国的情境，它是我们讨论面对现实的基础；其次本书试图简要地回顾中国的和西方的展览空间的历史和类型。接下来，我们抓住的是美术馆中的两个基本概念：观看与行动，通过艺术史的梳理我们才能建立起美术馆设计的前提。按照实际项目发生的需求，在不过多涉及建筑师主观个性的情况下，本书将演示如何将各种美术馆的（基本的和创新性的）功能诉求，一步一步地具体化与结构化，最终转化为成功建筑设计的雏形。至此，我们并不能告诉读者他们所希望看到的美术馆是什么样子的，但是他们至少对这个话题揭示的可能性有了更深入的理解。

将多样化的美术馆需求转化为僵化的设计标准是没有必要的，也不大可能，很难说，本书罗列的事项和观点会不会互相冲突。每个问题的提出可能基于不同的文化假设，所以不能指望本书中的观点成一大封闭的"体系"；要知道美术馆的服务对象千变万化，照顾了"古典"，就有可能忽略了"摩登"。更重要的是创造性往往来源于"意外"，过分追求清晰和理性，美术馆设计这件事儿也许就不好玩儿了。

作为一本全面覆盖美术馆设计议题的参考书，本书并不想把对具体问题的答案整合成一整套万应灵药式的解决方案。对那些有志于美术馆建筑设计的人士，它只想建立起适合当下实际的设计方法论，一种有益于实践的"演习地图"，它是一堂美术馆设计入门课的"教学大纲"——希望这本书的出版能够唤起专业人士对于本专题的注意，中国的建设大潮中涌现的美术馆，可以为这门课提供更具体鲜活的本土案例。

....

My Philosophy

Tang Keyang

What kind of design guidebook would Chinese art museum architecture need? This book is not meant to be one of the time-saver series of "architecture technology look-up booklet". Rather, this guidebook's main task is to articulate, dissect, and clarify the key issues facing museum designing. During the course of architectural practice, it is experienced that a good starting goal could be far more important than the means employed to achieve the goal. Why do we build an art museum? How is an art museum used and utilized? How can we maximize its utility under existing conditions in China? We need to bear all these in mind before commencing a museum design project.

Museum designing is not a purely technical job. Although the museum's art and culture implications cannot be easily concluded, nonetheless consensus and basis are needed. This book attempts to open up discussions, while trying to avoid idiosyncratic topics. In this book, we have established four categories of factors for discussion in the following order: urban context, art, architecture, and technique, with no implication of importance among the factors. Since there are hardly museum theories and principles to guide the ever-booming Chinese museum practice, we start with tackling the "starting goal" issues facing museum design. Our focus is to make readers aware of certain important topics, without the impractical expectations of generating a world-class answer.

However, it does not mean the guidebook lack standpoints. Our standpoint of analysis is based on the study of art itself. What is art, then? Unlike other cultural categories, contemporary art practice seeks to identify problems and "open" discussions, rather than "concluding" them once for all. This is particularly true of Chinese art in its modern form, a practice not yet clearly "defined" or accomplished, after continuous development over the past two centuries. This is true of Chinese art museums as well, the design of which should leave more room of future development for the ever-evolving Chinese art. Nowadays, museum designs tend to focus simply on exterior shapes and interior styles. Even when creativity happens accidentally, there is still a lack of theoretical basis from architects. In such unprepared and unguided situation, however, Chinese museums have mushroomed in a "Great Leap Forward" mode, where museum architecture does not understand art and the art circle does not appreciate architecture, while neither art nor architecture could satisfy the urban needs, not to mention numerous technical errors. Huge risks and regrets emanating from such circumstances are unimaginable.

It is in such a situation that this guidebook is written and compiled. The discussions are no longer arranged as in an encyclopedia or like journal entries in accounting books. Rather, we try to establish a reasonable framework with layered analyses. First, we begin with considerations about the immediate Chinese situation which constitutes the context of our discussion. Secondly, we briefly review the history and typologies of exhibition spaces from East to West. Thirdly, we analyze the two fundamental concepts in the museum – viewing and acting, with the reference in art history, in order to bridge art and architecture for museum designing. Fourth, we dissect various museum functions (basic and creative) and illustrate step by step how to configure and actualize the functions in the museum spatial system till their successful transformation into a museum design prototype – these discussions are based on the objective needs of real-life projects, without dealing with the subjectivity of architects. At this point, the guidebook still does not provide an answer to the readers of how a desired museum looks like. But it at lease makes them understand deeper the potentials in the theoretical perspective of this topic.

It is unnecessary and impossible to compile all diversified museum needs into a set of rigid standards of design, as some of the discussed topics and perspectives may turn out to be unresolvable paradoxes. These points and arguments are not meant to be a single, enclosed "system" of everything either, as questions are raised from different cultural assumptions. After all, museums serve a broad audience – for instance, focusing on "classics" may result in neglecting "moderns". More importantly, innovation usually happens from "small accidents" – pursuing undue rationality and clarity in museum design could probably kill the fun.

As a reference book covering a full range of museum design topics, this guidebook does not intend to compile answers to specific questions into a rigid set of solutions, like a panacea that kills all diseases. Rather, it intends to establish a design methodology to address the immediate reality, and offers a "mapping" of intellectual preparation for museum and architecture practitioners. This book sees itself as a syllabus to the "Introduction to Chinese Museum Design" course. Finally, we hope this book will bring attention to museum professionals and architecture practitioners; and hence new museums emerging from the rapid wave of Chinese urbanization and construction will contribute fresh local cases for further discussion.

附录

Appendix

六大美术馆基本信息比较 *
Case Studies of Six Museums*

	法国 巴黎 奥赛博物馆 Musée d'Orsay Paris, France	法国 巴黎 阿拉伯世界研究中心 The Arab World Institute Paris France	法国 巴黎 盖·布朗利博物馆 Quai Branly Museum Paris France
建成时间 Building Time	-1900	1981–1987	1999–2006
改扩建时间 Time of Extension	-1986		
新建时间 Time of New Building			
基本陈列（面积指标 / 所占比例） Permanent Installation (area/percentage)	16,500 m² 27.05%		9,250 m² 11.47%
临时陈列（面积指标 / 所占比例） Temporary Display (area/percentage)			2,000 m² 2.48%
藏品库（面积指标 / 所占比例） Storage (area/percentage)			9,250 m² 11.47%
公共服务（面积指标 / 所占比例） Public Service (area/percentage)			
专业技术（面积指标 / 所占比例） Technical Support (area/percentage)			5,000 m² 6.20%
物业（面积指标 / 所占比例） Building Management (area/percentage)			
设备（面积指标 / 所占比例） Equipment (area/percentage)			
行政（面积指标 / 所占比例） Administration (area/percentage)			
停车（面积指标 / 所占比例） Parking (area/percentage)			
交通（面积指标 / 所占比例） Transportation (area/percentage)			
结构（面积指标 / 所占比例） Structure (area/percentage)	14,000 m² 22.95%	8,351 m² 33.06%	
使用面积（面积指标 / 所占比例） Usable Area (area/percentage)	47,000 m² 77.05%	16,912 m² 66.94%	44,561 m² 55.24%
总建筑面积（面积指标 / 所占比例） Total Area (area/percentage)	61,000 m² 100%	25,263 m² 100%	80,674 m² 100%

* 数据可能依不同信源有所差异。
* Data may vary due to different sources.

	法国 巴黎 蓬皮杜中心 Le Centre Pompidou Sdulv/#udqf h	卡塔尔 多哈 卡塔尔国家博物馆 Qatar National Museum Doha, Qatar	阿联酋 阿布扎比 阿布扎比卢浮宫 The Louvre Abu Dhabi, UAE
建成时间 Building Time	1971–1977	1981–1987	1999–2006
改扩建时间 Time of Extension	1996–1999		2009–2014
新建时间 Time of New Building			2007–2014
基本陈列（面积指标 / 所占比例） Permanent Installation (area/percentage)	11,885 m² 11.50%	8,000 m² 15.66%	6,681 m² 10.45%
临时陈列（面积指标 / 所占比例） Temporary Display (area/percentage)		2,000 m² 3.91%	2,364 m² 3.70%
藏品库（面积指标 / 所占比例） Storage (area/percentage)	5,324 m² 5.15%	3,904 m² 7.63%	4,416 m² 6.91%
公共服务（面积指标 / 所占比例） Public Service (area/percentage)	10,888 m² 10.54%	16,936 m² 33.15%	5,611 m² 8.78%
专业技术（面积指标 / 所占比例） Technical Support (area/percentage)	5,681 m² 5.50%	7,736 15.12%	
物业（面积指标 / 所占比例） Building Management (area/percentage)			
设备（面积指标 / 所占比例） Equipment (area/percentage)			
行政（面积指标 / 所占比例） Administration (area/percentage)	3,591 m²	1,852 m² 3.62%	3,444 m² 5.39%
停车（面积指标 / 所占比例） Parking (area/percentage)	10,677 m² 10.34%	25,140 m²	1,735 m² 2.71%
交通（面积指标 / 所占比例） Transportation (area/percentage)		4,982 m²	8,783 m² 13.74%
结构（面积指标 / 所占比例） Structure (area/percentage)	14,684 m² 14.21%		
使用面积（面积指标 / 所占比例） Usable Area (area/percentage)	88,621 m² 85.79%		22,432 m² 35.08%
总建筑面积（面积指标 / 所占比例） Total Area (area/percentage)	103,305 m² 100%	51,167 m² 100%	63,942 m² 100%

国家美术馆新馆（提议）建筑竞赛第一阶段任务书摘要

1.1 应征设计方案应充分体现国家美术馆新馆项目"此时、此地"的要求。"此时"指的是国家美术馆在中国当代文化建设中继往开来的角色，"此地"指的是国家美术馆在北京奥林匹克中心区"首都文化之岭"的核心位置。

1.2 国家美术馆新馆在文化上应定位为兼有博物馆、美术馆和研究机构的文化设施，在物理形态上应定义为大型综合型博物馆。它的主要功能是展示、收藏、保管、修复和推广现当代艺术作品，发起、筹备、执行和宣传国家级和公共性的艺术项目。

1.3 应征设计方案应妥善考虑国家美术馆将来展出大量不同类型艺术品的可能性。"大量"是指展览的个体规模和同时举办展览的数目；"不同"是指展示对象和展览方式的多样性，尤其需要考虑富有中国特色的艺术品特有的展览方式。

1.4 应征设计方案应合理应对国家美术馆使用中的通达性问题。通达性问题包括但不限于单个和多个展览的流线组织，展品和日常供应的装卸和运输通道、办公和来访人员的进出和来往等，尤其需要考虑在现有条件下国家美术馆内部交通和区域城市交通的合理对接。

1.5 应征设计方案应恰当平衡建筑形象和功能之间的关系。国家美术馆的设计既要为首都北京乃至中国留下一座地标性建筑，也需要满足公共文化设施的基本功能，尤其鼓励创造性地处理建筑的外部形象和内部功能之间的关系。

1.6 应征设计方案应结合中国国情考虑建筑设计经济性的问题。经济性是指在同等条件下如何以最低的代价获取同样质量的设计，理想的设计方案应考虑到施工和维护的费用和基于本地的可操作性。

1.7 应征设计方案应完备考虑建筑单体和周边景观的有机融合。除了绿色设计的一般性原则，国家美术馆的环境设计应充分研究本地的气候条件和植栽特色，并将室外空间合理有机地融入国家美术馆使用的程序之中。

1.8 应征设计方案应长远计划国家美术馆作为未来北京重要公共空间的功能。这一功能既包括美术馆的现有功能，也包括具有文化前瞻性和具有中国社会特色的功能。

1.9 应征设计方案应考虑空间不断改造更新和灵活利用的可能性，尤其是指与新技术所带来的展览与建筑使用方式有关的变化，如基础设施改造、信息网络更新与电子展示系统的更换。

Guideline Excerpts from the First Round Competition of the Proposed NAMOC New Building

1.1 The bidding design should fully consider the temporal and spatial contexts in which the National Art Museum of China (NAMOC) project has been commissioned. The "temporal" context means that NAMOC should play the role to carry the historical mission of Chinese art and culture into future development while the "spatial" context refers to NAMOC's pivotal location at the capital city's cultural district, the Olympics Park.

1.2 The new venue of NAMOC, in cultural positioning, should combine the functions of a museum, an art gallery, and a research institute; it's physical form should be a large-scale comprehensive museum. Its primary functions include display, preservation, storage, restoration, and promotion of modern and contemporary art. It should be capable of initiation, preparation, execution, and publicization of state-level art programs open to the public.

1.3 The bidding design should take into consideration that the museum needs the capacity to handle large quantities of art works in various formats. "Large quantities" refers to the number and scale of exhibitions held at the same time; "various formats" refers to the diverse types of artworks and their varied ways of display, especially art forms uniquely Chinese in nature that require special ways to display.

1.4 The bidding design should take into consideration the accessibility of NAMOC during daily operations. "Accessibility" aspects include, but are not limited to, single and multiple exhibition routes, loading and moving of art objects and daily supplies, entrances and exits for visitors and staff, and in particular, points of connection between the museum and nearby transportation networks under current conditions.

1.5 The bidding design should carefully strike a balance between the functionality and external image. The design should make NAMOC a landmark building for Beijing and China, as well as functioning well as a public cultural facility. Architects are particularly encouraged to address creatively the relationship between the "external" image of the building and the "internal" functions of the building.

1.6 The bidding design should consider an economical way of designing under China's current conditions. The economical way means how to achieve the same quality of design at minimum cost. The ideal design should consider costs of construction and maintenance, as well as local operability.

1.7 The bidding design should blend organically the building with the surrounding cityscape. In addition to the general principles of a "green building", the environmental design for NAMOC should research on local climate and vegetation, and integrate outdoor spaces organically into NAMOC's operation program.

1.8 The bidding design should center on the functions of NAMOC as key public cultural space in Beijing in the future. Besides the current museum functions, the design should reflect forward-looking cultural functions with Chinese characteristics.

1.9 The design should consider the possibilities of flexible use of spaces and future upgrade of spaces, in particular, changes in exhibition and building utilization pertaining to technological innovation, such as infrastructure renovation, information network upgrade, and replacement of electronic display system.

中国美术馆设计介绍

戴念慈

首都的中国美术馆胜利地落成了。今年 5、6 月间，文化部和中国美术家协会为纪念毛主席《在延安文艺座谈会上的讲话》发表 20 周年，在这里举办了全国美术展览会。

美术馆的建造，主要是为了经常展出我国自五四运动以来在美术方面的各种优秀作品，包括绘画、雕塑、版画、工艺美术品和民间艺术品等；同时，将来国内或国外来我国的临时性美术展览会，也可以在这里举行；并有一部分面积辟作中国美术家协会办公和美术家们集会交谊等活动之用。全馆建筑面积约 16000 平方米，其中展出用的面积有 7000 平方米。目前，主体工程已经基本完成，仅馆外的庭园绿化工程和馆内某些装修和设备工程，还有待以后逐步实现。

建筑内容和平面布局

根据使用的性质来分析，美术馆的内容大致可以分为下列几个比较主要的部分：

1. 展览部分：包括大小展厅共 17 个，这是美术馆中最为重要的部分。
2. 供应服务部分：包括装帧、裱画、展品修补和木工车间，还有贮藏展品和贮藏各种器材的仓库等，它们是为展览部分服务的一个必不可少的部分，是展览厅的"后台"。
3. 办公部分：包括管理用房和美术家协会的办公用房等。
4. 公共活动部分：包括美术家们集会用的大厅——美术家之家——和国际交谊活动用的接待厅、贵宾休息室等。
5. 其他：包括通风机房、变电所等附属房屋。

以上这些部分，各自有其不同的使用特点，因此在平面布局中就必须根据这些特点来安排。例如，展览厅是直接提供广大观众观赏美术品的场所，是美术馆最主要的部分，因此都布置在整个平面中最显要的位置，以便于观众参观。其次，如接待厅和"美术家之家"等公共活动的场所，也适当布置在显要位置上。办公部分则和全馆其他各部分既要有一定的联系，又需要自成一个"天地"，以避免观众的干扰，这里就把它集中布置在建筑物的西端，有西边大门可供出入。供应部分布置在全馆的后部，连接后院，并有后门为器材出入之用。展品仓库即分设在各展览层和夹层里面。地下室主要作为通风机房，一小部分用作器材仓库。

以上是从内部的使用要求来分析平面布局的情况。从基地周围的环境来分析，建筑物南边临猪市大街，这是横贯北京东西向的一条主干道；东临王府大街，也是一条很繁华的干道。因此，在平面布局上，这两个面都应当加以适当的重视。由于这一情况，设计中把南面正中作为建筑物的主要入口，东西两边各设次要入口。从南边主要入口进门，有一个作为交通中枢的中央大厅，从这里直通中间的综合展览厅、侧展厅和楼上各展厅。从东西入口进门，即各侧门厅直达侧展厅和四角展厅（西门兼作美术家协会办公的出入口）。各展厅除内部互相联系外，正门入口的两边还有长廊从外面直接沟通正门和角展厅的联系。东西两边和三层的廊子则作为观众休息之用〔图1〕。

这样一个既分散而又有联系的平面布置，不但比较切合基地环境的特点，而且也给使用上带来了很大的便利。因为这个美术馆除经常性展览外，同时还常常有各种临时性的展览会，这些临时性展览会如果各自有适当的出入口，就可以避免内部过多的穿行人流，使展览厅内能有较好的秩序。如此，我们可以设想，将来某些观众也许进南面正门去参观经常性的展览，另外一些观众也许直接进东门去参观例如"齐白石逝世十周年纪念展览会"，还有的也许从正门外长廊直穿到西边角厅去参观"某某画家旅行写生展览会"。如果还有些人愿意浏览一下全部展览品，他也可以通过内部各展厅之间的联系，纵览全馆。看累了，可以随时到廊子里憩息片刻，看看水池里的睡莲和游鱼，回味一下刚才看到的优美的艺术品。三层楼上的观众，还可以到外边回廊上凭栏远眺，南望金碧辉煌的故宫和雄伟的人民大会堂，西看青翠蓊郁的景山和秀美的北海白塔，北面是古老的钟鼓楼，东面是新建的工人体育场、工人体育馆和远处成片的工业区。回到馆里，又可以继

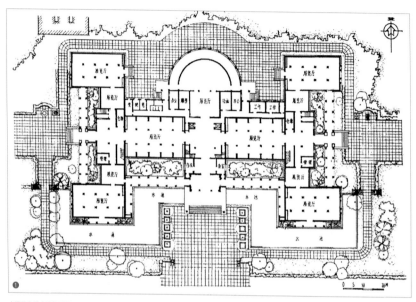

中国国家美术馆首层平面图
Plan of ground floor of National Art Museum of China

续欣赏那琳琅满目的美术作品。

采光问题

美术馆采光条件的好坏，常常直接影响到展品的展出效果。一幅色彩丰富、层次分明的油画，如果处在光线很差的地方，就会显得暗淡无光。因此，采光问题是美术馆设计中的重要课题。

在设计过程中，清华大学土木建筑系对这个美术馆的采光问题，作了很多研究工作和模型试验工作。目前这个馆的采光设计，基本上就是以清华方案作为蓝本的。从这次全国美术展览会的展出效果来看，采光处理一般还可以满足使用上的要求。这里也把这一问题简要地介绍一下。

富于中国传统风格特色的细节
Details featuring Chinese traditional Characteristics

美术馆的采光设计，一般要求解决几个问题：

1. 适度和均匀的照度。
2. 避免眩光。
3. 避免一次反射和二次反射——这两种反射光常常是一般美术馆中比较普遍存在而难于解决的问题。
4. 避免阳光直晒展品，使展品变质。
5. 采光口不占用或少占用展览墙面。

根据上述要求，一般建筑常用的侧窗采光或高侧窗采光方式就不能适用于美术馆的采光需要，因为这种做法虽在构造上比较简单，但是窗子占用了展览墙面，室内光线不匀，眩光、一次反射和二次反射都十分严重。太阳光直射展品的问题也不能完全避免^{（图4—5）}。

如果完全采用人工照明的方法，虽然比较容易根据展出的意图来控制光线，满足各种不同的要求，上述问题也比较好解决，但是这个办法设备投资和日常维持费用较大，而且灯光和太阳光在光质上亦存在一定的差别，因此，这种办法亦不完全相宜。

中国美术馆主要采用了顶部采光和侧窗采光两种形式。凡有条件做顶部采光的展厅，如四角展厅和三层展厅，都采用了顶部采光的办法，并在天窗下加折光以防止太阳直射。其他展厅则采用了高侧窗采光。但是由于展览厅进深宽度较大，为了避免眩光和反射光，采取了一些相应的措施，在高侧窗下面，加上一排折光片，使阳光从折光片折射到画面上去。这样，避免了眩光和直接的一次反射光，厅内的光线分布也比较均匀合理，近墙面处较强，中间站人处较暗，减少了二次反射的强度。虽然有些地方远处折光片对镜框面的反光仍有一定影响，但布置时适当调整画面高度和斜度，或者观众稍稍调整一下站立位置，就可以避免。从这次展出的实际效果看，情况也正是这样。

但是，这个馆的采光设计也还存在另外一些问题。其中最主要的问题是：虽然每个展厅本身的光线分布都比较均匀合理，而这个展厅和那个展厅之间的照度却很不一致。例如顶部采光的展厅就比侧窗采光的展厅亮很多。而且随着气候、季节和时间的不同，照度情况亦大有区别，如多云的天气太阳光由于天空漫射的作用，上述差别就比较小，晴朗无云的天气则差别比较大；同样顶部采光的展厅，冬季南边展厅光线较强，北边展厅的光线由于建筑物阴影的关系，就大大减弱；同一个侧窗采光的展厅里，冬季南墙的照度较大，而北墙则较小；而这些差别在夏天则并不显著。还有，同一天之内，早晨、中午和下午由于日照角度不同，建筑物阴影的位置推移变化，各展厅照度情况亦随着有所变动。当然采用了天然光来解决采光要求，总是要受到自然界各种变化的影响的。上述这些问题，借助于窗帘和人工照明，例如太亮的地方用窗帘遮挡一下，太暗的地方则用人工照明补助一下，是可以适当解决的^{（图4—5）}。

建筑形式问题

对于中国美术馆应该采取什么样的建筑形式，曾经由美术家协会的负责同志、有关领导、美术家、建筑师们多次研究讨论，提出了很多意见。根据我们的体会，这些意见大致可以归纳为以下几点：

普通侧窗（平面）
Side-window (plan)

普通高侧窗（剖面）
High side-window (section)

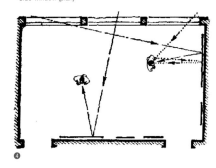

1. 要鲜明地表现中国的民族风格。
2. 要反映新中国美术创作上繁荣的气象，在形式上既要丰富多彩，又不能给人以豪华浮夸的感觉。
3. 要和附近的故宫、景山、北海等建筑群适当地协调呼应。
4. 要适当地处理好由于展览厅特殊要求而出现的大片光墙面。

设计中考虑了以上的意见，适当利用了我国传统的民族形式，在中间突出的四层部分（"美术家之家"）采用了中国古典阁楼式屋顶，其他部分都做成平顶以解决展厅的顶部采光问题，而在正门廊及各别几处休息用的廊榭则采用几个小小的中国式屋顶，略作点缀，以和中间顶部的楼阁相互呼应；再配上回廊和浅浅的琉璃檐子，就把整个建筑物在风格上结成一个整体了。这样做法，采用古典形式的屋顶的部分约占全部屋顶面积的十分之一多一点（平屋面女儿墙上的檐子不算），而整个建筑物的民族风味却烘托得相当浓厚[图 2, 3, 6, 7]。

在表面处理上，整个建筑物用一色浅米黄的陶制面砖作为贴面材料。装饰部分也都是陶制花饰，其中有的是琉璃面，有的是素陶面，两者间杂使用，没有采用大点金彩画或过于鲜艳的色彩。这样，既使整个立面显得比较丰富，也适当避免了豪华的感觉。但从目前实际效果来看，如果花饰的琉璃部分再少一些，把它们适当换成素陶面，琉璃花饰的颜色再淡雅一些，在效果上可能还会更好。至于内部处理，各展厅都没有作过多的装饰，色彩

美术馆主要立面的渲染图
Rendering of the building's main façades

也比较单纯，这是为了不使建筑环境过多地影响展览品、分散观众对展品的注意力。这样，展品的展出效果就会更加突出。

美术馆在形式处理上，也还有不少缺点，例如三层回廊柱子的排列，在构造上就不是很理想的。这是由于三层回廊在施工中途才决定增加，当时下层有些地方已经做好，在柱子安排上无法作通盘考虑，因而形成的一个缺陷^(图8—9)。

330

还有一个问题也是大家非常关心的，那就是美术馆和东南角上的华侨大厦、东边新建的大楼相互之间的协调问题。这三个成犄角之势的建筑物的不同体型，从目前来看，确有不够协调之处，现在有关部门正在研究解决。

总之，设计一个美术馆，大至建筑总体的关系，平面布局，采光通风，建筑形式，小至一个挂画方法的问题，都应当考虑周全，才能满足要求。这里不能一一叙述，只能就主要方面大致介绍如上，希望读者批评指正。

编者按：
戴念慈先生是 1960 年代中国美术馆建筑的总建筑师。本文摘选自《建筑学报》1963 年 08 期。本书对原文格式做了适当编辑处理，并增补由美术馆提供的插图。

Introduction to the Design of the National Art Museum of China in 1963

Dai Nianci

Construction of the National Art Museum of China, located in the capital, has been successfully completed. During May and June of this year at the museum, the Ministry of Culture and the China Artists Association (CAA) held National Art Exhibition to commemorate the twentieth anniversary of Chairman Mao's Talks at the Yan'an Forum on Literature and Art.

The National Art Museum is established to primarily cater to the needs of exhibiting exemplary artworks since May-Fourth Movement, including drawings, sculptures, prints, crafts and folk art. At the same time, the museum will host temporary exhibitions, both international and domestic, as well as providing offices for the CAA and spaces for art events and artist gatherings. The entire art museum covers roughly 16,000 square meters, with 7,000 square meters of it for exhibition uses. As of now, the majority of construction tasks are complete, except museum gardening and certain indoor furnishing and equipment installation to be finished.

The Categories and Layout of the Building

Based on functionalities, the spaces of the National Art Museum can be generally classified in the following categories:

1. Exhibitions: this category includes 17 galleries of difference sizes. This is the most important category of the museum.
2. Supply and Services: this category includes workshops for artwork framing, mounting, conservation and carpentry, as well as warehouses for artworks and exhibition equipment story. They provide indispensable services to exhibitions, and act as the "backstage" for all galleries.
3. Offices: this category includes museum management offices and offices for the CAA.
4. Public Spaces: this category includes a gathering hall for artists, known

as the Home of Artists, as well as reception lounges for international events and VIP rooms.

5. Others: this category includes ventilation rooms, substations, and other maintenance rooms.

The above categories all have their own functions and usages, and therefore must be arranged accordingly in the layout. For example, galleries, by providing the general public with exhibitions, are the most important category of the art museum. Therefore, galleries should occupy the principal areas in a building's layout, allowing easier access to the public. Public spaces, such as reception areas and the Home of Artists, should also be featured prominently in the layout. Offices should operate in a detached location to avoid disruption by the public, but still keep connected with other departments of the museum. Thus, the office functions are arranged in the west section of the museum building, with an entrance through the west gate. The supplies department goes to the back of the museum, connected to the back yard, with the back gate for transportation of equipment. The exhibition storages are located on the exhibition floors and mezzanine levels. The basement floors primarily store the ventilation system, with a small section used for equipment storage.

The above is the analysis of the layout in consideration of the functional requirements inside the building. An analysis of the environment surrounding the museum shows that the building faces in the south Zhushi Street, a major latitudinal road in Beijing, and in the east Wangfu Street, another prominent bustling street. Therefore, with regards to the layout of the museum, special attentions must be paid to these two façades. Accordingly, the main entrance is arranged at the middle of the south façade, and secondary entrances are placed on the east and west façades. Upon entering the main entrance from the south, visitors find themselves in a central hall, a visitor hub that leads to the main gallery in the center and to galleries on either side as well as upstairs. The east and west entrances lead to the side galleries and the four corner galleries, with the west gate serving also as the entrance for the CAA offices. The galleries are interconnected, with a long external veranda connecting the main entrance on both sides with the corner galleries. The verandas at the east and west sides and on the third floor serve as resting places (Figure 1).

Such a separated yet inter-connected layout not only matches the environment of the museum, but also gives great convenience to users. Besides conventional exhibitions, the art museum also hosts temporary exhibitions of various kinds. As such, offering separate entrances to temporary exhibitions avoids unnecessary foot-traffic inside the museum and maintains order within the galleries. We can imagine that within the space of one day, some visitors will enter from the main entrance in the south to visit conventional exhibitions, some may enter from the east entrance to visit the Memorial Exhibition of the Tenth Anniversary of the Legacy of Qi Baishi, while others may enter through the external west veranda to visit the west corner gallery for the exhibition "Drawings from Life in the Travel of An Artist". If some wish to view all the exhibits, they may tour the whole museum via the inter-connecting paths between galleries. When they get tired, they may take a break in the veranda at any time, enjoy the sight of the water lilies and fish in the pond, and savor the aftertaste of artworks they just saw. Visitors to the third floor may go to external veranda to enjoy views, overlooking to the south the magnificent Forbidden City and the Great Hall of the People, to the west the green Jingshan Mountain and the beautiful Beihai Pagoda, to the north the ancient Bell and Drum Tower, and to the east the newly-built Workers' Stadium, Workers' Gymnasium and the industrial zones beyond. Then, back into the museum from veranda, they can continue to appreciate the excellent artworks.

Lighting

The lighting conditions of an art museum can often directly affect the display of exhibits. An oil painting rich in color and layers may look dull if set in a dimly lit place. Thus, lighting is an important issue in the design of art museums.

During the design process, the School of Architecture at Tsinghua University did a lot of research and model testing of museum lighting; the current lighting of the museum uses the Tsinghua design as a blue print. Judging from the display effects of the recent National Art Exhibition, the lighting of the museum, generally speaking, can meet the requirements for usage. Here is a short summary of the issues considered.

Usually, the following problems need to be tackled in the lighting design of an art museum:

1. Appropriate and even luminance.
2. Avoidance of glare.
3. Avoidance of direct reflections and double reflections; both forms of reflective light are universally challenging problems in art museums.
4. Avoidance of direct exposure of the exhibits to sunlight.
5. Windows occupying none or very small wall spaces.

Judged by the above-mentioned requirements, side-window and high side-window lighting is not appropriate for the lighting needs of an art museum.Although side-window and high side-windows are structurally simple, all these – windows occupying wall spaces, uneven interior lighting, glare through windows, direct reflection and double reflection – pose serious problems. And direct exposure to sunlight could not be avoided fully (Figures 4 – 5).

Should full artificial lighting be adopted, control of light and capacity to meet different lighting requirements according to exhibition needs are much easier; it is easier to solve the above problems, too. Yet the cost of such equipment investment and daily maintenance will be higher, and there are quality differences between artificial light and sunlight. Thus, this approach does not completely work either.

Two lighting approaches are adopted at NAMOC, namely, skylights and side-window lights. In locations that allow for the installation of skylights, such as in the four corner galleries and the third-floor galleries, sky lighting is used. Reflective devices have been added under the sky windows to avoid direct exposure to sunlight.

High side-window lighting is adopted in other galleries. However, as the galleries have a relatively large depth, certain measures have been taken to avoid glare and reflection. An array of reflective devices is added below the high side-windows, which refract sunlight towards the paintings on the wall; in this way, glare and direct reflections are avoided and the luminance in the galleries are evenly distributed; with stronger light close to the walls and dimmer light towards the middle where visitors stand, double reflections of light are reduced. Although the reflective devices still affect the reflections on the framed surfaces, this problem can be avoided by simply adjusting the height and angle of paintings, or by adjusting the standing point of viewers. These solutions have actually worked as supposed in the recent exhibition.

Nevertheless, some other problems remain in the lighting design of the museum, most outstanding of which is the difference luminance in different galleries despite the even distribution of light within each gallery. For instance, galleries with skylights are much brighter than galleries with high side-window lights. With the changes of climate, season and time, the luminance varies greatly. For instance, the above difference is reduced in cloudy weather due to sunlight being diffused, but is more apparent on fine and cloud-free days. In galleries with similar skylights, in winter the light is stronger in the south-side galleries but weaker in the north-side galleries, due to the shadows of the building. In the same gallery with side-window light, in winter the luminance is stronger in the southern wall but weaker in the northern wall. Yet the aforementioned differences are less noticeable in summer. Also, in the same day, due to different angles of the sunshine in the morning, at noon and in the afternoon, building shadow positions shift, and thus, the luminance in different galleries change accordingly.

Of course, the use of natural light as museum lighting will be unavoidably affected by the changes in natural conditions. The above problems can be solved by way of curtains and artificial lighting: curtains can block some light where it is too bright, and artificial lighting could be used to supplement natural lighting where it is too dark (Figures 4 – 5).

Architectural Style

What kind of architectural style NAMOC should take is a question that has been discussed many times by comrades of the CAA in charge, relevant leaders, fine artists and architects, and many opinions have been proposed, which we have summarized as follows:

1. It must distinctly represent China's national style;
2. It must reflect the prosperous development of fine arts in New China, and formally it must be rich and colorful but without the feeling of pomposity and superficiality;

3. It must coordinate properly and harmoniously with the architectural groupings of the Forbidden City, Jingshan and Beihai;
4. It must properly deal with the issue of large tracts of blank walls as a result of the special display demands of the galleries.

In consideration of the above opinions, traditional Chinese forms have been properly employed. In the protruding part of the fourth floor (the Home of Artists), a classical Chinese-pavilion type of roof is adopted, while flat roofs are used elsewhere as a solution to sky lighting issues. Several small-size Chinese-style roofs are also applied to embellish the front porch and a few other verandas for resting, which echo the pavilion style in the middle top section. These, together with the verandas and the short eaves of glazed tiles, make the building integrated in style. With this practice, the classic-style roofs account for a little more than 10 percent of the whole roof area (the eaves on the parapet of the flat roofs not being taken into account), but the edifice as a whole emits indeed a strong national flavor (Figures 2, 3, 6, 7).

With regards to the façade treatment, the same light beige ceramic tiles are used throughout the whole building. The decoration also uses ceramic floral patterns, employing a mix of glazed tiles and plain ceramic tiles. No bright, gilded decoration or shiny colors are used. Thus, the façades have a varied appearance while avoiding the feeling of pomposity. However, judged by the current construction results, it would be better if part of the glazed floral tiles were replaced with plain ceramic tiles, and the colors of glazed floral tiles were lighter and more elegant. Regarding the interior treatment, no excessive decoration is used in the galleries, and the colors are simple; such treatment is employed to reduce the influence of architectural environment on the exhibits and to avoid distractions on the part of visitors, thus highlighting the artworks being displayed.

There are still some flaws in the treatment of the museum style. For instance, the arrangement of the third-floor veranda pillars is not very ideal in structural terms, as a result of the fact that the third-floor veranda was added later during the construction, when part of the lower level was already built and no thorough considerations could be given to the pillar arrangement – thus resulting in such a flaw (Figures 8 – 9).

There is another issue that has attracted people's attention, that is, the visual coordination with Huaqiao Plaza in the southeast and the new building in the east. The three buildings of different shape and form indeed are not well coordinated visually. This issue is being researched by the departments concerned.

In conclusion, in the design of an art museum, every detail – as big as general architectural composition, building layout, lighting and ventilation, architectural style, and as minute as the mounting of the paintings – must be carefully considered in order to meet requirements. As this paper cannot go into every detail, we only highlight major aspects of the design issues. Criticism and comments from the readers are welcome.

*The Editor:
Dai Nianci is the principal architect of the 1960s NAMOC project. The article is excerpted from The Journal of Architecture, vol.8, 1963, Beijing. The editor adjusted the format of the original article with more illustrations by courtesy of the museum.

美术馆设计功能推导表（第一组）
Sample Working Sheets for Museum Program (Group 1)

通行
Entrance

项目 Project	总面积 Total Area	独立 Independent	面积 Area	非独立 Non-independent	面积 Area
公共入口 Public Entrance		地铁和地下商场入口 Subway/Shopping Mall Entrance		通用入口 General Entrance	
		地下停车入口 Parking Entrance		团体入口 Group Entrance	
		残疾人入口 For the disabled		来客入口 Visitor Entrance	
办公入口 Office Entrance		办公入口 Office Entrance			
货品入口 Object Entrance		藏品入口 Collection Entrance		展品入口 Exhibition Object Entrance	
		物资入口 Non-Art Object Entrance			
		后勤入口 / 垃圾出口 Service Entrance			
公共聚会入口 Public Events Entrance				公共聚会入口 Public Events Entrance	
公共教育入口 Public Education Entrance				公共教育入口 Public Education Entrance	

公共信息与接待
Information & Reception

项目 Project	总面积 Total Area	独立 Independent	面积 Area	非独立 Non-independent	面积 Area
大厅 Lobby		接待厅 Reception Hall		入口大厅 Main Lobby	
信息 Information				票务 Ticketing	
				问询 General Inquiry	
				查询 Information Terminal	
访客服务 Visitor Service				前台 Front Desk	
				寄存 Bag Check	
安检处 Security Check				安检处 Security Check	
公共交流区（动） Public Meeting				公共交流区（动） Public Meeting	
公共休息区（静） Public Resting				公共休息区（静） Public Resting	
VIP 接待 VIP Reception		VIP 接待 VIP Reception			
儿童中心 Children Center				儿童中心 Children Center	

餐饮与零售
Dining & Retailing

项目 Project	总面积 Total Area	独立 Independent	面积 Area	非独立 Non-independent	面积 Area
零售 / 便利店 Grocery/Convenience Store				零售 / 便利店 Grocery/Convenience Store	
艺术商店 Art Store				纪念品商店 Souvenir/Book Store	
				艺术品商店 Art Store	
博物馆购物中心 Museum Shopping Center				博物馆购物中心 Museum Shopping Center	
商店储藏 Store Storage		商店储藏 Store Storage			
餐厅 Restaurant				中式餐厅 Chinese Restaurant	
				西式餐厅 Non-Chinese Restaurant	
快餐店 Dining Hall				快餐店 Dining Hall	
酒吧 / 咖啡 / 茶座 Bar/Café/Tea Shop				酒吧 / 咖啡 / 茶座 Bar/Café/Tea Shop	
厨房 Kitchen		厨房 Kitchen			

公共交流
Public Exchange

项目 Project	总面积 Total Area	独立 Independent	面积 Area	非独立 Non-independent	面积 Area
大厅 Auditorium Lobby				大厅 Auditorium Lobby	
多功能厅 Multi-use Hall				多功能厅 Multi-use Hall	
多媒体中心 / 音像室 Multimedia/Video-Audio				多媒体中心 / 音像室 Multimedia/Video-Audio	
影院 Theatre				影院 Theatre	
礼堂 Auditorium		休息室 Auditorium Resting Area		大厅 Auditorium Lobby	
		后台 Back-Stage			
讲座厅 Lecture Hall		讲座厅 Lecture Hall			
观众休息厅 Auditorium Reception				观众休息厅 Auditorium Reception	

公共教育
Public Education

项目 Project	总面积 Total Area	独立 Independent	面积 Area	非独立 Non-independent	面积 Area
大厅 Auditorium Lobby				大厅 Auditorium Lobby	
教育办公室 Public Education Office		教育办公室 Public Education Office			
图书阅览 Library		开架图书馆 Library Stack			
		闭架图书馆 Library Circulation			
		善本图书馆 Rare Collection			
		电子阅览室 Reading Room			
		小阅览室 Reserved Reading Room			
教室 Classroom		教室 Classroom			
		电脑教室 Computer Classroom			
		多功能教室 Multi-use Classroom			
美术室 Drawing Room		美术室 Drawing Room			
训练室 Training Room		训练室 Training Room			
工作室 Workshop		工作室 Workshop			
教育 / 董事会 Public Education Board Office				教育 / 董事会 Public Education Board Office	
学术研究 Research Room				学术研究 Research Room	
学术交流 Academic Exchange				学术交流 Academic Exchange	
学生中心 Student Center		学生中心 Student Center			

展览
Exhibition

项目 Project	总面积 Total Area	独立 Independent	面积 Area	非独立 Non-independent	面积 Area
永久展览 Permanent				现代美术精品编年史陈列 Permanent Display by Time	
				国际美术精品 Best of the International Art	
专门门类艺术展厅 Special Category		中国民间艺术陈列 Folk Art			
		中国书画展厅 Chinese Painting and Calligraphy			
当代艺术综合展区 Contemporary Art		专题展 Themed Exhibition			
				国际美术馆馆际交流 International Art	
				文化交流展厅 Cultural Exchange	
				学术交流展厅 Academic Exchange	
特殊展区 / 多功能厅 Special Gallery / Multi-use Hall		开幕式专用厅 Opening Area			
				大尺度艺术 Large-scale Artworks	
				开拓性实验性展览厅 Experimental Category	
				多媒体互动展区 Multimedia	
				表演艺术展厅 Performance Art	
艺术家驻场空间 / 艺术工作室 / 个人创作展示 Artist Studio/Workshop		艺术家驻场空间 Artist Studio			
		艺术工作室 / 个人创作展示 Workshop			
室外展场 Outdoor Gallery		室外展场 Outdoor Gallery			
周转区 Exhibition Setup Area					

藏品处理
Collection Processing

项目 Project	总面积 Total Area	独立 Independent	面积 Area	非独立 Non-independent	面积 Area
修复维护 Restoration and Conservation		消毒室 Cleaning		图画修复 2D Artworks	
				工艺美术修复 Crafts	
				文物修复 Cultural Relics	
				书画修复 Chinese Media	
				装裱室 Mounting	
				藏品维护 Conservation	
				新媒体维护 New Media	
藏品研究 Research		研究室 Research Room			
		鉴定室 Inspection Room			
工作室 Workshop				工作室 Workshop	
研究观摩 Research Gallery		研究观摩 Research Gallery			
实验室 Laboratory		实验室 Laboratory			
媒体资料中心 Media Center		藏品摄影 Photography/Dark Room			
		资料室 Archive			
		图像服务 Image Service			

收发
Receiving

项目 Project	总面积 Total Area	独立 Independent	面积 Area	非独立 Non-independent	面积 Area
装卸区 Loading Area				卸货平台 Loading Dock	
				搬运拆装区 Transportation and Box Opening	
				藏品装卸 Collection Loading	
				展品装卸 Exhibition Loading	
				物资装卸 Non-Art-Objects Loading	
收发 Receiving		艺术品入库登记 Art Objects Register		艺术品转运 Art Objects Transfer	
		接收包装 Art Objects Packaging		开箱鉴定 Art Objects Box Opening	
		收发接待室 Receiving Office		艺术品配送通道 Art Objects Distribution	
				艺术品装置区 Art Objects Setup	
停车场 Parking		VIP 停车场 VIP Parking		大巴停车 Bus Parking	
				办公专用停车 Office Parking	
武警办公室 Armed Police Office		武警办公室 Armed Police Office			
消防车道 Fire Emergency Pathway		消防车道 Fire Emergency Pathway			

室外
Outdoor Space

项目 Project	总面积 Total Area	独立 Independent	面积 Area	非独立 Non-independent	面积 Area
景观绿化 Landscaping		园林绿化区域 Greeneries		空中花园 Sky Garden	
				水景 Water Features	
				室内庭院 In-building Courtyard	
				室外广场 Open Plaza	
室外设施 Open Plaza				室外设施 Open Plaza	
露天建筑空间 Outdoor Facility				屋顶平台 Roof Deck	
				室外平台 Porch	
				露天过道 Outdoor Concourse	
未来扩展空间 Space for future Expansion				未来扩展空间 Space for future Expansion	

技术设施
Facilities/Structures

项目 Project	总面积 Total Area	独立 Independent	面积 Area	非独立 Non-independent	面积 Area
动力房 Power Room		动力房 Power Room			
机械室 Mechanic Room		机械室 Mechanic Room			
设备间 Equipment Room		设备间 Equipment Room			
洗衣房 Laundry		洗衣房 Laundry			

贮藏
Storage

项目 Project	总面积 Total Area	独立 Independent	面积 Area	非独立 Non-independent	面积 Area
2D 存储 2D Storage		纸上艺术储藏室 Paper			
		民间艺术储藏室 Folk Art			
		油画储藏室 Oil Painting			
		印刷储藏室 Prints			
		摄影储藏室 Photography			
3D 储藏 3D Storage		雕塑藏品室（金属） Sculpture (Metal)			
		雕塑藏品室（石头） Sculpture (Stone)			
		雕塑藏品室（木） Sculpture (Wood)			
				综合装置艺术藏品室 Mixed Media	
				大件作品藏品室 Large Scale	
珍品藏品库 Rare Collection		珍品藏品库 Rare Collection			
图书馆库房 Library Storage		图书馆库房 Library Storage			
临时存储 Temporary Storage for Art Object				暂存库 Temporary Storage	
				机动储藏 Reserved Storage	
开放式仓库 Open Storage				开放式仓库 Open Storage	
非艺术品仓库 Non-Object Art				非艺术品仓库 Non-Object Art	
工具储藏 Tools				展具储藏室 Exhibition Storage	
				器材储藏室 Equipment Storage	
教育储藏 Education Storage				教育储藏 Education Storage	
办公储藏 Office Storage				办公储藏 Office Storage	

办公
Office

项目 Project	总面积 Total Area	独立 Independent	面积 Area	非独立 Non-independent	面积 Area
办公门厅 / 接待 Office Lobby/Reception				办公门厅 / 接待 Office Lobby/Reception	
馆办公室 Office		办公室 Office		会议室 Conference Room	
		会客室 Guest Room			
		放映间 Screening Audio- Video			
		视听室 Video/Audio Room			
党委办公室 Party Committee's Office		党委办公室 Party Committee's Office			
馆长办公室 Director's Office		馆长办公室 Director's Office			
研究与策划 Research and Curatorial Affairs Department		研究与策划 Research and Curatorial Affairs Department			
人事部 Office Management Department		人事部 Office Management Department			
财务部 Financial Department		财务部 Financial Department			
展览部 Exhibition Department		展览部 Exhibition Department			
典藏部 Collection Department		典藏部 Collection Department			
收藏部 Acquisition Department		收藏部 Acquisition Department			
民间美术部 Folk Art Department		民间美术部 Folk Art Department			
公共教育部 Public Education Department		公共教育部 Public Education Department			
公共关系 / 对外联络 Public Relations Department/ International Department		公共关系 / 对外联络 Public Relations Department/ International Department			
信息传播 Information Department		信息传播 Information Department			
医务室 Clinic		医务室 Clinic			
档案室 Archive		档案室 Archive			
办公储藏 / 存储 Office Storage				办公储藏 / 存储 Office Storage	
后勤部 Service Department		后勤部 Service Department			
安保 Security Department		安保 Security Department			
清洁区 / 服务间 / 废物收集 Cleaning Service		清洁区 / 服务间 / 废物收集 Cleaning Service			
休息间 / 更衣间 / 淋浴间 Restroom/Change Room/Bath		休息间 / 更衣间 / 淋浴间 Restroom/Change Room/Bath			
员工餐厅 Dining Room				员工餐厅 Dining Room	
办公厨房 Kitchen				办公厨房 Kitchen	
办公庭院 Office Yard				办公庭院 Office Yard	

美术馆设计功能推导表（第二组）
Sample Working Sheets for Museum Program (Group 2)

通行
Museum Entrances

教育 Education	面积 Area	总面积 Total Area	娱乐 Recreation	面积 Area	总面积 Total Area	商业 Commerce	面积 Area	总面积 Total Area
公共教育入口 Public Education Entrance			通用入口 General Entrance			地铁和地下商场入口 Subway/Shopping Mall Entrance		
			团体入口 Group Entrance			地下停车入口 Parking Entrance		
			来客入口 Visitor Entrance					
			公共聚会入口 Public Events Entrance					

公共信息与接待
Information & Reception

教育 Education	面积 Area	总面积 Total Area	娱乐 Recreation	面积 Area	总面积 Total Area	商业 Commerce	面积 Area	总面积 Total Area
问询 General Inquiry			入口大厅 Main Lobby			票务 Ticketing		
查询 Information Terminal			公共交流区（动） Public Meeting			寄存 Bags		
前台 Front Desk			公共休息区（静） Public Resting					
安检处 Security Check			儿童中心 Children Center					
			接待厅 Reception Hall					

餐饮与接待
Dinning & Retailing

教育 Education	面积 Area	总面积 Total Area	娱乐 Recreation	面积 Area	总面积 Total Area	商业 Commerce	面积 Area	总面积 Total Area
						零售 / 便利店 Grocery/Convenience Store		
						纪念品商店 Souvenir/Book Store		
						艺术品商店 Art Store		
						博物馆购物中心 Museum Shopping Center		
						中式餐厅 Chinese Restaurant		
						西式餐厅 Non-Chinese Restaurant		
						快餐店 Dining Hall		
						酒吧 / 咖啡 / 茶座 Bar/Café/Tea Shop		
						商店储藏 Store Storage		
						厨房 Kitchen		

公共聚会
Public Gathering

教育 Education	面积 Area	总面积 Total Area	娱乐 Recreation	面积 Area	总面积 Total Area	商业 Commerce	面积 Area	总面积 Total Area
多功能厅 Multi-use Hall			大厅 Auditorium Lobby					
多媒体中心 / 音像室 Multimedia/Video-Audio			休息室 Auditorium Resting Area					
影院 Theatre			观众休息厅 Auditorium Reception					
后台 Back-Stage								
讲座厅 Lecture Hall								

公共教育
Public Education

教育 Education	面积 Area	总面积 Total Area	娱乐 Recreation	面积 Area	总面积 Total Area	商业 Commerce	面积 Area	总面积 Total Area
开架图书馆 Library Stack			大厅 Lobby for Public Education					
闭架图书馆 Library Circulation								
善本图书馆 Rare Collection								
电子阅览室 E-Reading Room								
小阅览室 Reserved Reading Room								
教室 Classroom								
电脑教室 Computer Classroom								
多功能教室 Multi-use Classroom								
美术室 Drawing Room								
训练室 Training Room								
工作室 Workshop								
学生中心 Student Center								

展览
Exhibition

教育 Education	面积 Area	总面积 Total Area	娱乐 Recreation	面积 Area	总面积 Total Area	商业 Commerce	面积 Area	总面积 Total Area
现代美术精品编年史陈列 Permanent Display			开幕式专用厅 Opening Area					
国际美术精品 International Art								
中国民间艺术陈列 Folk Art								
中国书画展厅 Chinese Painting and Calligraphy								
专题展 Themed Exhibition								
国际美术馆馆际交流 International Art								
文化交流展厅 Cultural Exchange								
学术交流展厅 Academic Exchange								
大尺度艺术 Large-scale Artworks								
开拓性实验性展览厅 Experimental Categories								
多媒体互动展区 Multimedia								
表演艺术展厅 Performance Art								
艺术家驻场空间 / 艺术工作 室 / 个人创作展示 Artist Studio/Workshop								
室外展场 Outdoor Gallery								

收发
Receiving

教育 Education	面积 Area	总面积 Total Area	娱乐 Recreation	面积 Area	总面积 Total Area	商业 Commerce	面积 Area	总面积 Total Area
						大巴停车 Bus Parking		
						办公专用停车 Office Parking		
						VIP 停车场 VIP Parking		

室外
Outdoor Space

教育 Education	面积 Area	总面积 Total Area	娱乐 Recreation	面积 Area	总面积 Total Area	商业 Commerce	面积 Area	总面积 Total Area
			空中花园 Sky Garden					
			水景 Water Features					
			室内庭院 In-building Courtyard					
			室外广场 Open Plaza					
			室外设施 Outdoor Facilities					
			屋顶平台 Roof Deck					
			室外平台 Porch					
			露天过道 Outdoor Concourse					
			未来扩展空间 Space for future Expansion					

世界美术馆索引
Index of World Museums

主题索引
Index by Topics

总策划
范迪安

编辑总监
谢小凡

编著
唐克扬

编辑组（按姓名汉语拼音不分先后）
付 海　韩德泉　姜 山　李宣谊
刘兰婷　刘 伟　邵菁菁　杨济瑜

英文翻译
郑 涛　Jacob Dreyer

英文审校
Liz Emrlch　Matthew Jellick　Lily Wang

设计
王学军

Project Director
Fan Di'an

Editorial Director
Xie Xiaofan

Editor/Writer
Tang Keyang

Editorial Team (by Last Name)
Fu Hai, Han Dequan, Jiang Shan, Li Xuanyi,
Liu Lanting, Liu Wei, Shao Jingjing, Yang Jiyu

Translation
Zheng Tao, Jacob Dreyer

English Proof Editor
Liz Emrich, Matthew Jellick, Lily Wang

Graphic Design
Wang Xuejun

图书在版编目（CIP）数据

美术馆指南 : 基于设计的角度 / 唐克扬编著. --
北京 : 生活·读书·新知三联书店, 2023.2
ISBN 978-7-108-06793-7

Ⅰ. ①美… Ⅱ. ①唐… Ⅲ. ①美术馆 – 建筑设计 – 研
究 – 中国 Ⅳ. ①TU242.5

中国版本图书馆CIP数据核字(2020)第014884号

策　　划	知行文化	
责任编辑	朱利国	
装帧设计	王学军	
责任印制	卢　岳	
出版发行	生活·讀書·新知 三联书店	
	（北京市东城区美术馆东街22号）	
网　　址	www.sdxjpc.com	
邮　　编	100010	
经　　销	新华书店	
印　　刷	北京隆昌伟业印刷有限公司	
版　　次	2023 年 2 月北京第 1 版	
	2023 年 2 月北京第 1 次印刷	
开　　本	850 毫米 × 1168 毫米 1/32　印张 11.25	
字　　数	160 千字 / 500 幅图	
印　　数	0, 001-5, 000 册	
定　　价	98.00元	

（印装查询：010-64002715；邮购查询：010-84010542）